Woody Biomass for Bioenergy Production

Woody Biomass for Bioenergy Production

Editor

Jaya Shankar Tumuluru

MDPI • Basel • Beijing • Wuhan • Barcelona • Belgrade • Manchester • Tokyo • Cluj • Tianjin

Editor
Jaya Shankar Tumuluru
Biological Processing Department,
Idaho National Laboratory
USA

Editorial Office
MDPI
St. Alban-Anlage 66
4052 Basel, Switzerland

This is a reprint of articles from the Special Issue published online in the open access journal *Energies* (ISSN 1996-1073) (available at: https://www.mdpi.com/journal/energies/special_issues/woody_biomass).

For citation purposes, cite each article independently as indicated on the article page online and as indicated below:

LastName, A.A.; LastName, B.B.; LastName, C.C. Article Title. *Journal Name* **Year**, *Volume Number*, Page Range.

ISBN 978-3-03943-993-5 (Hbk)
ISBN 978-3-03943-994-2 (PDF)

Cover image courtesy of Idaho National Laboratory, Idaho Falls, Idaho, USA.

Contents

Acknowledgments: This work was supported by the U.S. Department of Energy, Office of Energy Efficiency and Renewable Energy under DOE Idaho Operations Office Contract DE-AC07- 05ID14517. Accordingly, the U.S. Government retains and the publisher, by accepting the article for publication, acknowledges that the U.S. Government retains a nonexclusive, paid-up, irrevocable, worldwide license to publish or reproduce the published form of this manuscript or allow others to do so, for U.S. Government purposes.

The author would like to thank Dr. Fredrick Stewart, INL Biological Engineering Department Manager, and Dr. Vicki Thompson, INL Bioenergy Program Relationship Manager, for their support in getting this book published. Finally, the author would like to express extreme gratitude and thanks to his family for their help and support as well.

About the Editor

Jaya Shankar Tumuluru is the Principal Investigator of the "Biomass Size Reduction, Drying, and Densification" project funded by the US Department of Energy. His research at INL is focused on mechanical preprocessing and thermal pretreatments of biomass. He has published papers on biomass preprocessing and pretreatments; food processing, preservation, and storage; and prediction and optimization of the bioprocess using artificial neural networks and evolutionary algorithms. He has published more than 100 papers in journals and conference proceedings; presented more than 80 papers at national and international conferences; edited books on biomass preprocessing, valorization, and volume estimation; and published more than 15 book chapters. He was named an R&D 100 Award Finalist in 2018 and 2020 and received the INL Outstanding Engineering Achievement Award in 2019; the Asian American Engineer of the Year Award in 2018; the INL Outstanding Achievement in Scientific and Technical Publication Award in 2014; the Outstanding Reviewer Award for the American Society of Agricultural and Biological Engineers Food and Process Engineering Division in 2013, 2014, and 2016; the Outstanding Reviewer for the journal *Biomass and Bioenergy* in 2015; and the Outstanding Reviewer for the UK Institute of Chemical Engineers in 2011.

Preface to "Woody Biomass for Bioenergy Production"

The change in the climate policies can increase the use of biomass for bioenergy production. This can lead to greater consumption of woody and herbaceous biomass. Forests are expected to have an important role in climate change mitigation under future climate change policies. As an important renewable and sustainable energy resource, forest biomass is considered as the primary energy resource. In the United States, wood and wood-derived fuels contribute to about 2% of the energy consumed annually. According to the US Department of Energy, wood-based fuels will account for about 9% of the energy consumed by 2030.

Woody biomass can be converted to biofuels by different methods, such as thermal, chemical, and biochemical methods. As an energy source, woody biomass can either be used directly via combustion to produce heat or indirectly after converting it to different biofuels. Woody biomass can generate heat or electricity solely or in a combined heat and power (CHP) plant. Most of the bioenergy produced from woody biomass is consumed by pulp and paper mills. Also, woody biomass as pellets is widely used for heat and power generation. As an energy feedstock, woody biomass can be used with other energy sources, such as coal. The use of raw woody biomass for bioenergy production is a challenge due to physical properties and chemical composition limitations. Mechanical preprocessing technologies, such as size reduction and densification, and moisture management techniques, such as drying, help to address density and moisture issues. To improve the chemical composition, thermal pretreatment, such as torrefaction, is gaining significant importance as it increases the energy content and improves the proximate and ultimate composition, making woody biomass more suitable for cofiring pyrolysis and gasification in solid and liquid fuel production.

The focus of the book is to understand the preprocessing and pretreatments for forest biomass in bioenergy generation. The specific areas covered by this book are how forest biomass for biofuels impacts greenhouse gas emissions; mechanical preprocessing, such as densification of forest residue biomass, to improve physical properties such as size, shape, and density; the impact of thermal pretreatment temperatures on woody biomass chemical composition, physical properties, and microstructure for thermochemical conversions such as pyrolysis and gasification; the grindability of torrefied pellets; use of wood for gasification and as a filter for tar removal; and understanding the pyrolysis kinetics of biomass using thermogravimetric analyzers.

Jaya Shankar Tumuluru
Editor

 *energies*

Review

A Review on the Potential of Forest Biomass for Bioenergy in Australia

Sam Van Holsbeeck [1,*], **Mark Brown** [1], **Sanjeev Kumar Srivastava** [1,2] and **Mohammad Reza Ghaffariyan** [1]

[1] Forest Research Institute, University of the Sunshine Coast, Locked Bag 4, Maroochydore, QLD 4558, Australia; mbrown2@usc.edu.au (M.B.); ssrivast@usc.edu.au (S.K.S.); mghaffar@usc.edu.au (M.R.G.)
[2] School of Science and Engineering, University of the Sunshine Coast, Locked Bag 4, Maroochydore, QLD 4558, Australia
* Correspondence: svanhols@usc.edu.au; Tel.: +61-7-5456-5167

Received: 29 January 2020; Accepted: 2 March 2020; Published: 3 March 2020

Abstract: The use of forest biomass for bioenergy in Australia represents only 1% of total energy production but is being recognized for having the potential to deliver low-cost and low-emission, renewable energy solutions. This review addresses the potential of forest biomass for bioenergy production in Australia relative to the amount of biomass energy measures available for production, harvest and transport, conversion, distribution and emission. Thirty-Five Australian studies on forest biomass for bioenergy are reviewed and categorized under five hierarchical terms delimiting the level of assessment on the biomass potential. Most of these studies assess the amount of biomass at a production level using measures such as the allometric volume equation and form factor assumptions linked to forest inventory data or applied in-field weighing of samples to predict the theoretical potential of forest biomass across an area or region. However, when estimating the potential of forest biomass for bioenergy production, it is essential to consider the entire supply chain that includes many limitations and reductions on the recovery of the forest biomass from production in the field to distribution to the network. This review reiterated definitions for theoretical, available, technological, economic and environmental biomass potential and identified missing links between them in the Australian literature. There is a need for further research on the forest biomass potential to explore lower cost and lowest net emission solutions as a replacement to fossil resources for energy production in Australia but methods the could provide promising solutions are available and can be applied to address this gap.

Keywords: forest biomass; Australia; biomass energy potential; emission; bioenergy

---

1. Introduction

Forest biomass can provide additional revenue for forest managers and supply a bioenergy market to reach renewable energy targets. Using forest biomass for bioenergy has become an integrated part of forestry and a priority for all biomass utilization projects [1]. Large quantities of forest biomass are sustainably used around the world to generate heat, steam, and electricity through gasification and combustion processes [2,3]. Opposed to global bioenergy trends [4], there is little public or political support for the use of forest biomass in Australia [5]. With the lack of economic incentives, most of the non-merchantable forest harvest residues are burned in the forest or left to decompose on site. The bioenergy market represents only 4% of total energy production in Australia [6] and, of this, forest biomass is 25%, and bagasse is 29% [6,7]. Other renewables, including hydro (16%) and wind (12%) energy, have increased in the last decade [7]. Establishing a sustainable bioenergy market from biomass in Australia requires consideration of the availability of forest biomass, sustainable harvesting, the cost of the biomass supply chain (BSC) and the greenhouse gas (GHG) emissions related to bioenergy production. The BSC encompasses many technical, economic and environmental constraints associated with harvesting, handling, storage, transport and conversion facilities [2,8–10]. Satisfying both environmental and economic objectives is a significant consideration for establishing the bioenergy industry [11].

The potential of forest biomass is categorized in the literature [12] into four sequential terms according to calculations needed in the assessment of biomass for biomass energy potential. The theoretical biomass potential relates to the annual yield of forest biomass per unit of area and can be considered the upper-bound of the potential [12]. Restrictions introduced by alternative biomass uses and efficiency at a biomass collection level are included. The term is also modified in the literature [13] as biologically available biomass and includes a range of ecological and economical reductions of the initial biomass to determine what is usable. Alternatively, a New Zealand literature example [14] refers to a similar term as total recoverable residue volume. Generally, the theoretical biomass potential captures all restrictions at a stand or production level. Secondly, the available biomass potential describes the energy that technically and economically can be harvested and transported for energy purposes before conversion [12]. This includes some limitations related to harvest machinery, truck size and transport distance. The term captures measures and restrictions of biomass energy at a harvest and transport level. Next, the technological biomass potential is defined by the energy that can be produced bound to conversion technology, the capacity of the conversion facility and the efficiency [12]. The term applies to research focused on-site identification for potential power facilities when inputs around available biomass, facility capacity and technology, and maximum allowable transport distance are known. The technological biomass potential accounts for technical restrictions on the available biomass and efficiency of the technology at an energy conversion level. At last, the economical biomass potential is the part of the energy that is distributed with respect to competing energy sources [12]. The term includes the energy production cost and the capacity of the facility, or rather the profitability of the proposed investment. A whole range of cost estimates of the entire supply chain can be included to determine if the economical biomass potential is feasible at a distribution level. The term environmental biomass potential is added to the four sequential terms described in the literature [12]. The environmental biomass potential sits at the same level of analysis as the economical biomass potential but opens perspective in emissions and other environmental measures related to bioenergy production. The term includes emissions that occur during energy production and non-biogenic emissions due to the use of machinery that affects the carbon-neutrality of the bioenergy system. In similarity to its economic counterpart, it pays respect to competing for energy sources and evaluates bioenergy in comparison with a reference or fossil energy system.

Forest biomass is considered a sustainable source of energy; however, only when grown and harvested in a sustainable manner [15]. The sustainability of forest biomass production systems must consider that forest harvest residues help sustain the fertility of the site, regulate water flow and maintain plant, microbial and animal biodiversity [1,16,17]. These considerations are classified under the constraints of the theoretical biomass potential. Additionally, economic sustainability constraints such as fuel versus food, efficient energy balance, and social constraints determine what is or is not a sustainable resource. International forestry guidelines and forest certification ensure sustainable forest management but need to address the specific impact of the additional harvest of forest biomass [18]. In Australia, guidelines such as the Forest Stewardship Council (FSC) and Responsible Wood, are inclusive of forest biomass harvesting to the extent of encouraging the harvest and respecting environmental values [19,20]. However, none of these guidelines provide directions on environmental, economic and social concepts of the theoretical biomass potential in Australian forests.

The overall cost of the BSC includes the economics related to harvest, collecting, transport and conversion of forest biomass. The cost of an economically sustainable BSC is heavily influenced by operating costs and the need to maintain a supply of forest biomass [21,22]. The economics related to harvesting technology and collection methods [23] as well as biomass handling and storage are critical to ensure the reliability of supply [8,23]. The biggest cost contribution comes from transport which is determined by the quality and moisture content of the biomass and the mode of transport [24]. Forest biomass densities are generally low (400–900 kg/m^3) and moisture contents high (>50%), which results in transportation contributing 20–40% of the BSC cost [8,22,24–26]. Many of these cost measures determine the available biomass potential. Processing costs for converting forest biomass to bioenergy are determined by the technology used, the capacity of the plant, and the consistency/quality of supply [22]. In general, biochemical and thermochemical techniques are the most suitable for the conversion of forest biomass [27,28]. Processing into liquid or gaseous fuels can be done by biochemical conversion while combustion, gasification, and pyrolysis can be used to produce fuels, heat and electricity [11,27–30]. However possible, the forest biomass to the biofuel supply chain is still at a pre-commercial level in terms of technology [8]. These characteristics, together with the large, complex equipment required, and often the need for different transportation modes, create complex economic and logistic issues and result in losses of the technological biomass potential [8,22,31,32]. Several measures are in place to define the technological biomass potential. The input of energy, or primary energy that is already delimited by theoretical and available biomass potential constraints throughout the supply chain, and the type of technology and conversion efficiency, define how much useful energy we can get from the primary energy source. Different types of energy products are then sent to customers through the grid, networks or channels of distributors, wholesalers and retailers as net energy. In order to substitute for fossil fuel, it is important to check the energy balance of the proposed bioenergy system [33,34]. The net energy ratio, for example, is an indicative measure to ensure the system does not use more energy than it creates [33,35]. The ratio is defined by the produced energy/consumed energy ratio that equals the primary energy at the gate. The ratio is a supportive measure of the technological biomass potential and provides additional insight into the profitability of the system and thus the economical biomass potential.

The economic potential of the forest biomass supply is a function of the biomass availability and the profit during sequential steps of the value chain [8]. These measures aim to determine if the electricity production cost of a bioenergy facility is lower than the conventional power facility. Factors like internal return rate and net present value define profitability measures of the economical biomass potential and are indicators that allow us to accept or reject a proposed investment [12]. Several cost parameters should be considered to compare bioenergy and fossil energy systems. These include equipment and capital cost, the construction cost of power line and grid connection cost, stumpage cost for forest biomass, supply chain cost and additional maintenance and administrative cost [8]. Each of these costs contributes to the energy production cost or cash outflow in the net present value of the investment. The impact of the value chain can be extensive, and there is an unavoidable degree of uncertainty in the supply of forest biomass that makes the estimation of cost and profit hard to predict. Optimization and simulation models are tools that can provide further insight into the economical biomass potential, where the inclusion of a reference fossil fuel scenario should be considered [8].

Using forest biomass for bioenergy reduces GHG emissions compared to the use of fossil fuels and thus mitigates climate change. Biomass from a sustainably managed forest can be considered as a carbon-neutral energy source [36–38] since the carbon emitted during the energy conversion process is fixed relatively quickly during subsequent photosynthesis and tree-growth [37,39]. Life cycle assessment (LCA) of sustainably-sourced forest biomass for energy shows a period of climate warming impacts as a result of the delay for the CO_2 to be captured by new tree growth [36,40–42]. Evaluating environmental impact in an LCA must be comprehensive [22] and include all non-biogenic carbon emissions from the consumption of fossil fuels for production, transport, harvesting, collection, and pre-processing [43]. Other factors such as other atmospheric pollutants (e.g., methane) and the effects of direct and indirect land-use change affect the value chain and are important to the result of the LCA [11,43]. Land-use changes, due to replacing crops with intensive forest plantations, can increase GHG emission [11,44] but a change from crops that demand high fertilizer and pesticide inputs to a forest that produces biomass for bioenergy can reduce GHG emissions [45–47]. Including emissions as a measure of the environmental biomass potential in balance with energy cost as a measure of the economic biomass potential in value chain optimization is becoming increasingly important for the sustainable utilization of forest biomass [22].

This case study review aims to identify gaps and approaches used to assess the potential of forest biomass for bioenergy generation in Australia and is structured around the hierarchical nature of biomass potential as defined in [12]. Evaluating Australian studies on forest biomass for bioenergy in Australia in their methods used to determine the theoretical, available, technological, economical and environmental biomass potential. In the next section, we explain the scope and methods that define the extent of the literature review. The following five sections review how research achieved the respective levels of detail in forest biomass potential. We discuss distinctive features and limitations in such a way as to make recommendations with regard to measures of forest biomass potential.

2. Scope and Methods

This study reviews forest biomass that includes pulpwood, forest harvest residues (FHR), and sawmill residues. Terms like remaining slash or logging residues are considered FHR and are plant materials that remain on-site after conventional logging. Pulpwood is logs harvested for pulp and paper but is also suitable for bioenergy production, usually obtained through thinning practices. Sawmill residues for this study include offcuts, dust, and shavings. Other forms of forest biomass are small hardwood logs [48]; coarse woody debris (CWD) [49] and wood from mechanical fuel load reduction [50], which resemble either FHR or pulpwood. Studies on firewood for domestic use are not included in this study unless it is part of their recourse in combination with one of the previously described forest biomass types. Regrowth and short rotation trees have potential as biomass for energy generation, but there are many uncertainties with regard to their distribution and climatic tolerance [51–53] and are not discussed in this review.

Assessing the theoretical biomass potential for bioenergy generation requires measurement of above-ground biomass. Logs used for timber and paper industries are not included in this assessment. Attributes of the supply chain using forest biomass must be present to categorize studies at a harvesting, collection and transport (available biomass potential) and conversion (technological biomass potential) level. The economical biomass potential concerns cost related to the production and distribution of biomass energy with respect to alternative energy uses and the environmental biomass potential includes GHG emissions related to the conversion of forest biomass to bioenergy along the value chain.

Online research papers published in English language academic journals were obtained by searching electronic databases including Google Scholar, Scopus, and Web of Science. The keywords used in searches were: 'Australia' and 'forest' and a combination of 'energy', 'biomass', 'bioenergy', 'waste', 'residue', 'supply + chain', 'emission' and 'sustainable'. Review papers were included, but book chapters and reports were excluded. The literature search covered the period 2000 to 2020. For each paper, the following information was collected and analyzed:

1) Primary publication data: year of publication, author(s), affiliations and journal titles;
2) Abstract and keywords;
3) Presence of the five biomass potentials;
4) Measures and attributes included in the calculation of the theoretical, available, technological, economical or environmental biomass potential.

Thirty-five original research journal articles were identified that assessed the use of forest biomass in Australia for bioenergy. The majority of these were published in the journals Biomass and Bioenergy (9), Forest Ecology and Management (6), and Australian Forestry (5). Most research (31 out of 35) considered the theoretical biomass potential and sixteen studies considered the available biomass potential. In ten research studies, the theoretical biomass potential was combined with the available biomass potential. Technological, economical and environmental biomass potential studies are far less frequent with studies including a technological or environmental biomass potential being the lowest in frequency (6 out of 35). Figure 1 depicts the classification of the studies on forest biomass potential based on occurring measures in the research to provide an overview.

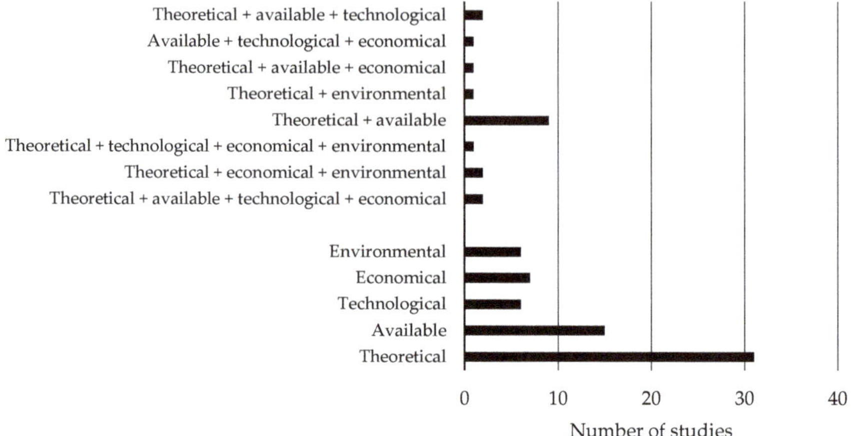

Figure 1. Classification of the reviewed studies on forest biomass for bioenergy in Australia according to the included measures of forest biomass potential.

3. Theoretical Biomass Potential

Thirty-one research studies include an assessment of the theoretical biomass potential (Table 1). In eighteen of these studies, the theoretical biomass potential is combined with the assessment of either available, technological, economical or environmental biomass potentials. Various studies have used allometric equations to predict the theoretical biomass potential based on in-field measures of height and diameter allowing them to estimate the overall above-ground biomass ratios [54–60]. For the purpose of bioenergy, these studies are rather informative literature on the ratio of above-ground biomass in order to assess the theoretical biomass potential without including any losses or alternative uses of the biomass. One can consider them indications of the upper-bound of biomass in the respective location and forest type. Similarly, an allometric equation is used to determine CWD removal benchmarks in the native forest of eastern Australia with the results being very specific for dead material only [49]. The study delivers good insight into the methods used to estimate the upper-bound of CWD but has no indication of further losses or potential of this biomass when extracted. Ximenes et al. (2006 and 2008) assessed the green weight of biomass using purpose-built trailers with built-in measuring devices [61,62]. The method provides an indication of the ratio of biomass in different forest types including some of the key tree species. The assessment of the theoretical biomass potential is once again restricted to a maximum allocation of the biomass without any indication of losses or harvestable volumes. Several papers calculated biomass quantity using the Cooperative Research Centre (CRC) for Forestry standard methodology [63] for sampling remaining slash [23,64–70]. In each of the papers, the CRC methods give a good indication of the number of residues that are left on-site before and after the removal of forest biomass. The methods have been tested in several case studies across Australia and have been applied in different harvesting systems like whole-tree, cut-to-length and integrated harvesting.

Table 1. Summary of forest biomass for bioenergy studies in Australia and indication on their inclusion of measures to assess the theoretical, available, technological, economic and environmental biomass potential.

Reference	Case	Theoretical	Available	Technological	Economical	Environmental
Fung et al. 2002	Outlining technologies for the conversion of woody biomass for heat and power generation.	×	×	×	×	
Specht and West 2003	Predicting biomass availability and carbon stock on farm forest plantations in northern New South Wales.	×				
Ritson and Sochacki 2003	Predicting biomass and carbon stock of *Pinus pinaster* trees in farm forestry plantations, southwest Australia	×				
Ximenes et al. 2006	Predicting the proportion of above-ground biomass in commercial logs and residues of spotted gum forest in southeast NSW.	×				
Raison 2006	Reviewing biomass supply technology, policy, availability and other impediments to expand forest bioenergy.	×			×	×
Cowie and Gardner 2007	Assessing GHG mitigation impacts of sawmill residues used either for the generation of electricity or the manufacture of particleboard.	×		×	×	×
Ximenes et al. 2008	Predicting the proportion of above-ground biomass in commercial logs and residues of five commercial forest species.	×				
Bi et al. 2010	Predicting above-ground biomass of *Pinus radiata* plantations from stand variables, geographical growth and yield models.	×				
Ghaffariyan et al. 2011	Evaluating biomass harvest in a poor-quality eucalypt plantation on yield and the productivity rates of equipment for harvesting.	×	×			
Rodriguez et al. 2011	Predicting the potential biomass availability for energy generation from forestry and agriculture in the Green Triangle.	×	×	×	×	
Farine et al. 2012	Predicting current and future biomass feedstocks for bioenergy, and associated estimates of the GHG mitigation.	×			×	×
Ximenes et al. 2012	Estimating GHG balance and emissions of two critical native forest areas managed for production in New South Wales in comparison with a conservation only scenario.	×				×
Ghaffariyan et al. 2012	Assessing biomass harvest using the Bruks mobile chipper for non-merchantable stem wood at the roadside in a pine plantation in Victoria.	×	×			
May et al. 2012	Assessing the energy balance of wood from softwood plantations and native hardwood forests, using a cradle-to-gate inventory.	×	×			
Moroni 2013	Reviewing GHG mitigation trade-off between storing carbon in forests and providing society with wood products.					×
England et al. 2013	Estimating GHG emissions associated with wood from softwood plantations and regrowth hardwood native forests.					×
Ghaffariyan 2013	Predicting remaining slash in different sites of plantations harvested by cut-to-length and whole tree method.	×				

Table 1. *Cont.*

Reference	Case	Theoretical	Available	Technological	Economical	Environmental
Ghaffariyan et al. 2013	Estimating BSC cost influenced by five operational factors: energy demand, moisture mass fraction, interest rate, transport distance, and truck payload.	×	×		×	
Walsh and Strandgard 2014	Predicting remaining slash after harvest stem wood biomass product from integrated cut-to-length harvest operations in *Pinus radiata* plantations.	×	×			
Mendham et al. 2014	Predicting the impact of repeated residue removal, retention, or retention of double the quantity of residues over two rotations of *Eucalyptus globulus* in southwest Australia.	×				
Ghaffariyan et al. 2014	Evaluating biomass harvest using a mobile chipper in a clear-felled area of pine plantations in Victoria.	×	×			
Meadows et al. 2014	Predicting the potential biomass for energy supply from hardwood plantations within the Sunshine Coast Council region of southeast Queensland.	×				
Ghaffariyan and Apolit 2015	Predicting remaining slash in Queensland pine plantations harvested by cut-to-length and whole tree method.	×				
Ghaffariyan et al. 2015	Evaluating biomass harvest of an integrated energy wood harvesting system compared to conventional log harvesting in a 32-year-old *Pinus radiata* plantation located in southwest Western Australia.	×	×			
Rothe et al. 2015	Predicting the current use and the potential sustainable supply of forest biomass in Tasmania.	×	×			
Cummins et al. 2016	Reviewing the biomass supply to produce liquid fuels and electricity using small hardwood logs.		×	×	×	
Crawford et al. 2016	Predicting spatial availability of biomass for bioenergy in Australia in 2010, 2030 and 2050.	×				
Wang et al. 2017	Predicting residue weight of individual trees in rotation age (28 to 42 years) *Pinus radiata* stands under three thinning regimes.	×				
Ximenes et al. 2017	Reviewing biomass harvest from mechanical fuel load reduction.	×	×	×		
Ghaffariyan et al. 2017	Reviewing biomass supply and recovery technologies from forests to gate.		×			
Ngugi et al. 2018	Predicting potential harvestable biomass for bioenergy from sustainably managed private native forests in southeast Queensland.	×				
Woo et al. 2018	Optimizing biomass energy facility locations, and allocation for supply.	×	×	×		
Garcia_Florez et al. 2019	Developing species-specific biomass estimation models (BEMs) for stem, bark, branch and crown compartments in 16-year old plantations of *Eucalyptus dunnii* and *Corymbia citriodora*.	×				
Threlfall et al. 2019	Predicting the amount of CWD in the harvested and unharvested native forest of eastern Australia.	×				
Strandgard et al. 2019	Reviewing BSC experience applied to reduce costs in emerging Australian forest BSC.	×	×			

Three studies assessed the theoretical potential of forest biomass across the nation using a range of literature assumptions applied on statistical data of the forest industry reported by the Australian Bureau of Agricultural and Resource Economics and Sciences (ABARE) [53,71,72]. Their research combines geographical distribution of the forest with annual forest production data and sawmill survey results. To estimate the quantity of biomass the research includes losses through a range of diversion and conversion parameters retrieved from the literature. The combination of statistical inventory data and literature is also applied in a method to estimate the theoretical forest biomass potential on the Sunshine Coast Council in a 20-year prediction [73]. Similarly, a current and potential forest biomass scenario was established for Tasmania [74]. The studies by Farine et al. (2012) and Crawford et al. (2016) also rely on the use of the 3-PG model [75], which is a process-based forest growth model, to estimate the theoretical biomass potential of forest biomass in the future. A similar method was applied in the Green Triangle in South Australia to assess the year-by-year biomass availability [76]. Because of the smaller scale and unified nature of the forest in the Tasmanian research paper [76], researchers were able to include more detail on forest biomass availability by including thinning practices and moisture content. Ximenes et al. (2012) used the Forest Resource and Management Evaluation System (FRAMES) model to predict the wood supply and converted yield to theoretical biomass potential of two case study areas [77]. The model includes three modules on inventory, growth and mortality models, and yield simulation including a future prediction for the next 200 years. Another case study in Tasmania [78] used a non-industrial private native forest (NIPNF) modelling approach to assess theoretical biomass potential in Tasmania. Their research includes a range of limitations on the land availability of harvest of forest biomass, year-to-year variation, moisture content and rotation of forest harvest. The importance of moisture content was also highlighted in a New South Wales case study [79] where three different scenarios of forest biomass are compared in their total emission and energy. Other than having the direct impact of moisture content on energy production, there were no other indications as to how respective biomass quantities were measured in the three cases [79].

The remaining studies review the theoretical potential of forest biomass but do this as part of a literature review of the potential of forest biomass for bioenergy in Australia or the application of biomass supply chain on a national scale [5,80,81]. Lastly, Ximenes et al. (2017) discuss some of the measures to assess the potential of CWD and standing forest biomass for mechanical fuel load reduction.

4. Available Biomass Potential

In fifteen of the reviewed studies, the available biomass potential is measured or discussed (Table 1). An international study [25] reviews some previous research in Australia on the cost related to harvesting and forwarding of biomass from forest to roadside, primarily using time-motion-studies, and reference some of the following work in different harvesting systems. Cost (USD/t), fuel consumption (L/t), and energy content (MJ/t) of slash bundling operations and total operational cost (USC/kWh) of a slash-bundler application to collect harvest residues in Eucalyptus plantation are measured using time-motion-studies [65]. The chipping cost (USD/t) and forwarding cost (USD/t) are analyzed using time-motion-studies on Bruks Chipper operations in Pine plantations [23,67]. Similarly, time-motion-studies in integrated harvest sites in Pine plantation were carried out to measure productivity and unit cost (AUD/m^3) of harvesting and forwarding operations [69]. Walsh and Strandgard (2014) also used time-motion-studies to assess the productivity of harvest and extraction in a Fibreplus operation in Pine plantations. None of these studies assess the available biomass potential after potential losses that occur during transport.

Rodriguez et al. (2011) used a transportation model to calculate the costs of transporting logs and chips. Literature values were used for the costs of collection and processing of biomass in the field. This model incorporated fixed and variable (costs to estimate the average costs per tonne-km^{-1} transport based on the one-way distance to destination and load mass. Fixed costs included capital depreciation, interest charges, labour, registration, insurance, repairs, maintenance, and salaries. Variable costs covered fuel, oil, and tyres. Using a 16–18 MJ/kg conversion, they estimated how much biomass energy

(MJ) was available before conversion. May et al. (2012) used the SimaPro model to estimate the energy use in forestry operations and included the establishment, management, harvesting and transport of logs. The analysis included calculations for transport, mean travel distance, load weight and fuel consumption and other materials. They presented an intermediate energy (MJ) value and used a 19.6 MJ/kg conversion rate for biomass chips. The BIOPLAN model [70] calculates the cost of the BSC and is based on the following factors: tonnes biomass, solid and lose volumes of biomass, truckloads, energy contents (17.38 MJ/kg), and costs of harvesting, extraction, chipping, storage and transportation. Woo et al. (2018) included the cost of transport and moisture content in a linear programming model to identify locations for new bioenergy facilities based on the available biomass potential. All of these modelling approaches include measures of the available biomass potential to the extent of what is technologically and economically harvestable.

Strandgard et al. (2019) reviewed the potential relevance to Australia of overseas supply chain models such as MCPLAN [24] and other measures of the available biomass potential (transport distance, load size, harvest rates and machine types). Ximenes et al. (2017) reviewed some considerations and conditions around different harvesting systems of CWD which result in the material either retained or captured for use. The remaining studies in Table 1 do not include any measures directly related to the available biomass potential other than the energy content of their respective resource and case study [48,74,77,79,81].

5. Technological Biomass Potential

The efficiency of conversion technology is determined as one of the most impacting factors of the technological biomass potential. The primary energy (often mentioned in MJ) going into conversion gets reduced according to the type of technology used, the output energy type (unit of Watt) and the efficiency of the process [81]. Efficiency indicates how much useful energy we can get from an energy source according to the conversion technology. Some of the Australian forest biomass research studies have included a sensitivity analysis to determine the efficiency of the conversion [48,79]. Variations in the efficiency of 10% in a combusting co-firing plant are tested with results reflecting on emission resulting from burning the biomass [79]. Variation (15–20%) in the efficiency of a Fisher–Tropsch synthesis is tested together with transportation distance to confirm if the supply of biomass supports the demand [48]. A conversion efficiency of 30% was used for a combined heat and power plant in the literature [77] and 25% for direct combustion by a Tasmanian research study [76] stating this lower-cost solution with smaller plant sizes (5 MW) tend to be more profitable in comparison with different biomass-based technologies for electricity generation. As expected, efficiency is the most common technological criteria to evaluate the bioenergy system. No further measures of energy ratio or energy balance were included.

Woo et al. (2018) determined the potential capacity of power facilities based on transport distance and moisture content by performing location-allocation network analysis. By supplying the required demand, they identify the most optimal locations to put power facilities with the least cost contribution from transport. The resulting capacity of the facility then relies on the amount of biomass that lies within the distance threshold and is economically transportable (available biomass potential). No conversion technology or efficiency was selected and no sensitivity analysis was performed. The methods can be used, however, for the design and planning of future facility locations and allocation of resources, providing that different capacities and technologies (including conversion efficiency) be tested to determine the lowest cost and emission solution.

6. Economical Biomass Potential

Several studies identify the economical biomass potential as an indicative value based on literature but do not include any measures of electricity production cost or return rates or the production of electricity of other energy types. Fung et al. (2002) mention a 3.4 TWh per year electricity potential from wood processing residues in Australia. According to Raison (2006), this potential lies 2.8 TWh per year based on dry forest biomass. Farine et al. (2012) estimate a 13 TWh energy potential and is the most recent published value based on forest biomass.

The only study in Australia effectively measuring the economical biomass potential compared to the electricity generation cost of biomass powered system with a coal-fired generation in a case study [76]. The research estimated the Levelized Electricity Cost (LEC) using a 30-year facility life and an annual interest rate r of 7.5%. The following formula was established:

$$\text{LEC} = \frac{\sum_{t=1}^{n} \frac{I_t + M_t + F_t}{(1+r)^t}}{\sum_{t=1}^{n} \frac{E_t}{(1+r)^t}}, \tag{1}$$

where I_t represents the initial investment for a direct combustion biomass electricity facility and M_t represents the maintenance and operating costs (excluding biomass purchase). The feedstock expenditure F_t is estimated based on the plant gate price of the feedstock multiplied by the amount of biomass required to run the plant, and finally, E_t is the amount of electricity generated every year.

There are no literature examples on optimization or simulations of the forest biomass supply or value chain that indicate electricity production cost for a bioenergy system in Australia. In the future, there could be the potential to establish costs associated with reductions of GHG emissions and costs associated with social benefits. Thus far, no studies in Australia were able to capture these costs in their methods.

7. Environmental Biomass Potential

The environmental biomass potential, more particularly the displaced GHG, and net GHG mitigation were assessed by Farine et al. (2012) who used forest inventory reports and literature conversion factors. This study calculated the emission (the equivalent of CO_2) associated with the various pathways from feedstock to bioenergy by summing the emissions related to production, transport, conversion, and combustion in Australia. Several scenarios of using forest biomass and the role of forest in GHG mitigation in Australia are reviewed [82] and implemented in a New South Wales case study [77]. The New South Wales case study used a carbon accounting model to estimate the greenhouse gas balance in a comparison between a timber-production and a conservation scenario. Modelling considered emissions resulting from the establishment and management of forests, harvesting, transport to the mill, manufacturing, transport to consumer and disposal. The research also compared results with emissions from the manufacture of non-wood products and the use of fossil energy. No specifics of the models' calculation are provided, and most of the emission factors are retrieved from a life cycle inventory report from Australia [83].

The SimaPro LCA model used by England et al. (2013) was combined with survey data on emissions and the Australasia Ecoinvent database and incorporated emissions from burning fossil fuel, non-CO_2, fire, establishment, management, harvest, haulage, transport, and fertiliser. A sensitivity analysis was used to assess the influence of varying parameters and assumptions on carbon emission and tested in a ±20% range. No details on the LCA model and calculations are provided.

Cowie and Gardner (2007) performed a comparative study on the use of sawmill residues for the generation of electricity and the manufacture of particleboard using emission equations. Calculations of GHG emissions included those associated with harvesting, collecting, chipping transport, processing at the mill, evaporation of moisture, and re-establishment of the plantation [79]. The study uses a reference fossil fuel scenario to calculate the net emission reductions from the two alternative bioenergy scenarios. In comparison, they excluded al common factors between the reference and case scenarios. For the bioenergy cases, the avoided emission is calculated considering the variation in carbon emission per unit primary energy and the variation in the efficiency of combustion. Emission of CO_2 that are avoided are calculated as the mathematical product of CO_2 equivalent in the biomass utilized and a "displacement factor" (DF), where:

$$DF = \frac{\text{efficiency of bioenergy system}}{\text{efficiency of fossil system}} \times \frac{CO_2 \text{ emission per J fossil fuel}}{CO_2 \text{ emission per J biomass}}. \tag{2}$$

Emissions of CO_2 as a result of biomass combustion are excluded based on biomass production from a sustainably harvested plantation.

8. Discussion

The review above demonstrates a number of methodological approaches as well as variations in data, data models, scale on which the methods are applied and the detail of calculations that are included. The purpose of this discussion is to elaborate on some of the main results and challenges that need to be addressed in future research.

Almost every study presents measures of the theoretical biomass potential to a certain degree. There is however a clear distinction between studies that do an initial measurement of forest biomass which could be referred to as the maximum or upper-bound biomass potential and studies that included a measure that reduces that potential to what will be considered usable as defined by Shi et al. (2008). Field measurements involving weighing biomass or using allometric equations based on tree height and diameter are complex and time-consuming but provide reliable estimates of that upper-bound biomass potential and provide forest or species-specific estimates. These case-specific values can then be used for large-scale or nationwide estimates of the forest biomass potential [53,71,72,84] that use these literature values with the modelled prediction on losses of the theoretical biomass potential. Inevitably, the accuracy of this combination decreases as research becomes more generalized. Using case-specific inventory data on the upper-bound biomass and generalizing this over larger, sometimes nationwide, areas give a doubtful indication of the theoretical biomass potential. It is, however, hard to say what is good or bad and often the level of accuracy is sufficient on a larger scale. This coarser resolution and large-scale are still relevant from a policy and planning point of view where it provides sufficient indication of biomass location and the possible potential. Alternatively, there are in-field measures of the usable biomass available like those referred to as the CRC for Forestry method described in the literature [63]. Methods like CRC for forestry, are applied after a timber harvest and give a better indication of the parts of the leftover biomass that could be collected for energy purposes, and reduce the level of assumption for the theoretical biomass potential. Sequentially, combining it with high accuracy growth models can predict future levels of biomass on a local scale, which then can be used as indicative measures of future potential, which will be used to establish renewable energy targets. Based on the available research around the theoretical biomass potential, there is no clear gap to identify. Future research should thus consider the level of accuracy that is required and the data that is available to guide the methodology. It is, however, worth noting that no method exists that measures only the amount of biomass that will be used for bioenergy and will be collected and transported. Even the CRC for Forestry method has an emphasis on "collectable" biomass; the results also indicate a lot of work goes into measuring biomass like cones, needles, leaves and twigs that are not feasible for collection. We suggest that this might require an approach working top-down by determining the biomass that is suitable for conversion in the facility, will have higher gate price, can be transported and harvested with high efficiency, to then determine what percentage of the total above-ground or post-timber harvest biomass this represents. To do this, the link between the theoretical biomass potential and the available, technological, economic and environmental biomass potential needs to be established in the Australian supply chain context. Another recommendation concerning the theoretical biomass potential is to use it as an instrument to inform policy in Australia and to adjust biomass-harvesting guidelines. Currently, none of the forestry guidelines including FSC and Responsible Wood is descriptive on the retention of forest biomass with respect to environmental, economic and social benefits.

Having been critical in some of the models used in the previous studies, it is worth recognizing that they provide direct insight into some of the planning issues regarding harvest and transport and thus the available biomass potential in relation to the theoretical biomass potential. More specifically, there is a particular emphasis being placed on the use of geographical data in combination with the literature or inventory data on the forest biomass. Some of the models that were used in the reviewed studies [70,71,76,78] make this exact connection as to how the cost of harvesting and transport affect the retention of biomass in the forest. Rather than looking at a steady-state situation of the forest, a method to assess the available biomass potential should look into changes in the biomass quantity over time as well as changes in the geographical location of biomass. Several studies can be found in the international literature, simulating changes in the supply of biomass-based on stand age, fire salvage or terrain [85–88]. If research wants to inform investors and if the location is pre-defined, high resolution of the geographical area and theoretical biomass potential should be combined to assess what is the available biomass potential. In addition to that, there is a need for performance values of the equipment, used along the supply chain. Although performed in just a handful of case studies in Australia, Ghaffariyan et al. (2017) summarize some of the operational costs of harvesting forest biomass equipment. For reference to transportation costs, some international models can be used as described by Strandgard et al. (2019). The assessment of the available biomass potential opens a whole new level of detail again, from different harvesting technologies for the timber as well as the forest biomass, different sizes and loads of trucks for transport, potential pre-treatment and storage to reduce cost and emission during the supply chain. Extensive modelling can be performed and numerous constraints can be added to the research model [2,9,22]. Although not as extensively researched as some studies in other countries or geographic areas, the biomass supply chain modelling research [70,71,76,78] that has been performed in cases in Australia is adequate to give an indication of the available biomass potential and to develop strategies in the commercial harvest of biomass for bioenergy production. Given the lack of commercial cases in Australia that effectively use forest biomass for bioenergy production, there is no established supply chain as a reference. Therefore, the word "optimisation" in research should be used carefully, as Australia aims to find a range of solutions rather than one lowest-cost or lowest emission solution. The type of optimization that can be performed is illustrated by Woo et al. (2018) where a plant location-optimization is performed based on the available resource and cost estimates of the supply chain. This type of research does address the biggest cost element of the supply chain which is transport [24] but is not affected by other elements, like harvesting or storage of the resource. Instead, the optimization is used as a planning and development tool to identify locations that are suitable for biomass conversion based on the availability of the resource, and thus paying respect to the theoretical and available biomass potential. Additionally, the network analysis of this study [78] delivers insight into the demand for energy which then leads into the technological and economical biomass potential. Similar approaches can be found in the international literature combining measures of the theoretical and available biomass potential [13,89,90]

In regards to the technological biomass potential the study by Woo et al. (2018), is perhaps the most comprehensive to find in Australia, even though no selection of conversion technology or sensitivity was applied. The method they use allows for additional measures of the technological biomass potential and has been used widely in the industry [13,90–92]. Network analysis based on spatial information can be used to identify the location for bioenergy conversion based on biomass availability and cost attributes of the supply chain. The assessment of the biomass potential of a regional network should be considered carefully when expanded to larger scales. Additionally, network analysis can be used to identify the size of the processing plant based on demand, technology and supply as well which is demonstrated in several case examples in New Zealand [14], Finland [93], Mozambique [94] and the USA [95,96]. The method is a streamlined combination of geographic information systems, mathematical modelling and network analysis that aligns in a decision support system. This is where some of the gaps in research become very apparent for Australia. The number of studies that assess the technological, economic and environmental biomass potential is significantly lower than the studies identifying the theoretical and available biomass potential discussed earlier (Figure 1). A couple of studies tested the sensitivity of an established biomass conversion technology [48,79], and do well in comparing the technology with a reference fossil conversion system which indicated gains of carbon emission and economic return over time. However, it is hard to draw conclusions from these two cases. It is imperative to have cases looking at specific conversion technology and efficiency to use in combination with planning and decision support systems to assess the large-scale technological potential of forest biomass. An example case study from the international literature [97], looks at the sensitivity of minimum energy recovery and thermal energy production constraints to impose that the amount of renewable energy produced meets the demand of the catchment area.

In addition to what Voivontas et al. (2001) introduced as the theoretical, available, technological and economical biomass potential, this review introduces the environmental biomass potential. One can assume the environmental biomass potential flows from the technological biomass potential in similarity with the economical biomass potential. Rather than having an economic output, the environmental biomass potential pays respect to carbon emissions resulting from the use of forest biomass for bioenergy. As argued in the literature [17,22], the use of forest biomass for bioenergy should consider economic and environmental constraints in order to assess its feasibility. Out of the three studies that had an economical and environmental element in their research [5,53,79], two studies were able to provide simultaneous measures of the displaced emissions and the provided energy [53,79]. Other than a potential energy production for distribution, there was no information on the cost of energy production, or cost-savings compared to a reference scenario as performed by Rodriguez et al. (2011). Whether performed separately or combined, one could argue that the use of LCA and value chain optimization provides a solution to address the potential of forest biomass from an economical and environmental point of view [8,98]. Australian studies by May et al. (2012) and England et al. (2013) estimated the embodied energy used and emissions from cradle to gate in the Australian forestry context. Their research emphasizes the difference between LCA as a means of assessing emission and energy associated with wood products including alternative uses, compared to life cycle inventory and cradle-to-gate inventory providing emissions associated with forest supply chain [84]. Thus, there is a need for a clear definition of the forest biomass for bioenergy system when it comes to assessing the economic and environmental value. Good research examples in Australia are available however not on the use of forest biomass for bioenergy. Several international studies can be found capturing constraints for the theoretical, available, technological, economic and environmental biomass potential [10,99,100]. A reference case study in Australia is research by Murphy et al. (2015) and Hayward et al. (2015) on the economics and sustainability of producing aviation fuels from energy crops in Queensland Australia. Their research indicates some of the future challenges of the industry, which include the expansion of case studies on a larger scale to make it reliable and sustainable. This challenge confirms findings in this review where we identify a need to use some of the case-specific data in combination with modelling to provide estimates on a larger or nationwide scale. A second challenge is to demonstrate that the industry can satisfy community demand and compels with sustainability expectations. Indeed, one of the findings of this review is to come up with a more comprehensive design and planning solutions for the forest biomass supply chain and the establishment of new conversion facilities. This is where the use of geospatial data and modelling comes in place to develop a decision support system that not only satisfies the supply but also lives-up to demand sustained energy production. Another recommendation is the need for full LCA to determine the total carbon footprint. However, defining the limit of LCA is important and can be challenging. Bioenergy projects in Australia are fairly new and thus the total effect on the energy market is yet to be discovered, especially in the long term. In addition, the adverse effect on the timber industry and other industries that deliver biomass feedstock needs to be considered. The last challenge that has relevance to the use of forest biomass for bioenergy is the risk of uncertainty of the supply chain and its elements. Future research has to bridge this gap mainly through securing the supply chain and finding a long-term supply of biomass for energy production. This finding enforces the need to make connections between the theoretical, available, technological, economic and environmental biomass potential once again.

9. Conclusions

In order to exploit biomass for bioenergy in Australia, the potential of forest biomass needs to be assessed, and methods need to be established to determine and evaluate the potential. This review applied four definitions of biomass potential as defined by Voivontas et al. (2001) and added a fifth definition for the environmental biomass potential. Although giving general definitions, this review evaluated some of the measures of biomass potential found in Australian forest biomass for bioenergy studies.

Almost every study includes measures to assess the theoretical forest biomass potential. The link with the available or technological, economical and environmental biomass potential is rare, however. Promising methods using decision support systems based on geographical data and modelling can be developed to make this connection and methods like LCA and value chain optimization might provide insight into the extensive possibilities around the economical and environmental biomass potential.

These methods have been used in other parts of the world and, with appropriate and accurate data, could be used in Australia to estimate costs-savings and emissions as an end-result. This information can then be used to initiate investment, simulate what can be achieved, and optimize business solutions to grow, harvest, transport and convert forest biomass for bioenergy.

Author Contributions: Conceptualization, S.V.H., M.B., S.K.S. and M.R.G.; methodology, S.V.H.; investigation, S.V.H.; data curation, S.V.H.; writing—original draft preparation, S.V.H.; writing—review and editing, M.B., S.K.S. and M.R.G.; visualization, S.V.H.; supervision, M.B., S.K.S. and M.R.G.; project administration, S.V.H. All authors have read and agreed to the published version of the manuscript.

Funding: This research was funded by the Australian Biomass for Bioenergy Assessment (ABBA) as part of the Australian Renewable Energy Agency (ARENA) through a University of the Sunshine Coast Research Scholarship (USCRS-ABBA), grant number (PRJ-010376).

Acknowledgments: We want to thank Prof Richard Burns, Dr Madaline Healey and the anonymous reviewers and the journal editor for taking their time to review and provide valuable input and comments. The Gottstein Trust provided additional funding through a forest industry top-up scholarship.

Conflicts of Interest: The authors declare no conflict of interest. The funders had no role in the design of the study; in the collection, analyses, or interpretation of data; in the writing of the manuscript, or in the decision to publish the results.

References

1. IEA Bioenergy. *Sustainable Production of Woody Biomass for Energy A Position Paper Prepared by IEA Bioenergy*; IEA Bioenergy: Rotorua, New Zealand, 2002; Volume 03.
2. Sharma, B.; Ingalls, R.G.; Jones, C.L.; Khanchi, A. Biomass supply chain design and analysis: Basis, overview, modeling, challenges, and future. *Renew. Sustain. Energy Rev.* **2013**, *24*, 608–627. [CrossRef]
3. Bridgwater, A.V.; Toft, A.J.; Brammer, J.G. A Techno-Economic Comparison of Power Production by Biomass Fast Pyrolysis with Gasification and Combustion. *Renew. Sustain. Energy Rev.* **2002**, *6*, 181–246. [CrossRef]
4. IEA. *Key World Energy Statistics*; International Energy Agency: Paris, France, 2017.
5. Raison, R.J. Opportunities and impediments to the expansion of forest bioenergy in Australia. *Biomass Bioenergy* **2006**, *30*, 1021–1024. [CrossRef]
6. KPMG. *Bioenergy State of the Nation Report*; Bioenergy Australia: Canberra, Australia, 2018.
7. Department of the Environment and Energy. *Australian Energy Update 2018*; Australian Government: Canberra, Australia, 2018.
8. Shabani, N.; Akhtari, S.; Sowlati, T. Value chain optimization of forest biomass for bioenergy production: A review. *Renew. Sustain. Energy Rev.* **2013**, *23*, 299–311. [CrossRef]
9. Malladi, K.T.; Sowlati, T. Biomass logistics: A review of important features, optimization modeling and the new trends. *Renew. Sustain. Energy Rev.* **2018**, *94*, 587–599. [CrossRef]
10. Cambero, C.; Sowlati, T.; Pavel, M. Economic and life cycle environmental optimization of forest-based biorefinery supply chains for bioenergy and biofuel production. *Chem. Eng. Res. Des.* **2016**, *107*, 218–235. [CrossRef]

11. Creutzig, F.; Ravindranath, N.H.; Berndes, G.; Bolwig, S.; Bright, R.; Cherubini, F.; Chum, H.; Corbera, E.; Delucchi, M.; Faaij, A.; et al. Bioenergy and climate change mitigation: An assessment. *GCB Bioenergy* **2015**, *7*, 916–944. [CrossRef]

12. Voivontas, D.; Assimacopoulos, D.; Koukios, E.G. Assessment of biomass potential for power production: A GIS based method. *Biomass Bioenergy* **2001**, *20*, 101–112. [CrossRef]

13. Shi, X.; Elmore, A.; Li, X.; Gorence, N.J.; Jin, H.; Zhang, X.; Wang, F. Using spatial information technologies to select sites for biomass power plants: A case study in Guangdong Province, China. *Biomass Bioenergy* **2008**, *32*, 35–43. [CrossRef]

14. Hock, B.K.; Blomqvist, L.; Hall, P.; Jack, M.; Möller, B.; Wakelin, S.J. Understanding forest-derived biomass supply with GIS modelling. *J. Spat. Sci.* **2012**, *57*, 213–232. [CrossRef]

15. Szulecka, J. Towards sustainablewood-based energy: Evaluation and strategies for mainstreaming sustainability in the sector. *Sustainability* **2019**, *11*, 493. [CrossRef]

16. Thiffault, E.; Hannam, K.D.; Paré, D.; Titus, B.D.; Hazlett, P.W.; Maynard, D.G.; Brais, S. Effects of forest biomass harvesting on soil productivity in boreal and temperate forests—A review. *Environ. Rev.* **2011**, *19*, 278–309. [CrossRef]

17. Evans, A.; Strezov, V.; Evans, T.J. Sustainability considerations for electricity generation from biomass. *Renew. Sustain. Energy Rev.* **2010**, *14*, 1419–1427. [CrossRef]

18. Stupak, I.; Lattimore, B.; Titus, B.D.; Tattersall Smith, C. Criteria and indicators for sustainable forest fuel production and harvesting: A review of current standards for sustainable forest management. *Biomass Bioenergy* **2011**, *35*, 3287–3308. [CrossRef]

19. FSC Australia. *The FSC National Forest Stewardship Standard of Australia*; FSC Australia: Melbourne, Australia, 2016.

20. Responsible Wood. *Australian Standard Sustainable Forest Management*; Responsible Wood: Brisbane, Australia, 2013.

21. Meyer, J.; Hobson, P.; Schultmann, F. The potential for centralised second generation hydrocarbons and ethanol production in the Australian sugar industry. In Proceedings of the 34th Annual Conference of the Australian Society of Sugar Cane Technologists, Palm Cove, Australia, 1–4 May 2012; Volume 34, pp. 585–596.

22. Cambero, C.; Sowlati, T. Assessment and optimization of forest biomass supply chains from economic, social and environmental perspectives—A review of literature. *Renew. Sustain. Energy Rev.* **2014**, *36*, 62–73. [CrossRef]

23. Ghaffariyan, M.R.; Sessions, J.; Brown, M. Evaluating productivity, cost, chip quality and biomass recovery for a mobile chipper in Australian roadside chipping operations. *J. For. Sci.* **2012**, *58*, 530–535. [CrossRef]

24. Acuna, M. Timber and biomass transport optimization: A review of planning issues, solution techniques and decision support tools. *Croat. J. For. Eng.* **2017**, *38*, 279–290.

25. Ghaffariyan, M.R.; Brown, M.; Acuna, M.; Sessions, J.; Gallagher, T.; Kühmaier, M.; Spinelli, R.; Visser, R.; Devlin, G.; Eliasson, L.; et al. An international review of the most productive and cost effective forest biomass recovery technologies and supply chains. *Renew. Sustain. Energy Rev.* **2017**, *74*, 145–158. [CrossRef]

26. Acuna, M.; Anttila, P.; Sikanen, L.; Prinz, R.; Asikainen, A. Predicting and controlling moisture content to optimise forest biomass logistics. *Croat. J. For. Eng.* **2012**, *33*, 225–238.

27. Yue, D.; You, F.; Snyder, S.W. Biomass-to-bioenergy and biofuel supply chain optimization: Overview, key issues and challenges. *Comput. Chem. Eng.* **2014**, *66*, 36–56. [CrossRef]

28. Lattimore, B.; Smith, C.T.; Titus, B.D.; Stupak, I.; Egnell, G. Environmental factors in woodfuel production: Opportunities, risks, and criteria and indicators for sustainable practices. *Biomass Bioenergy* **2009**, *33*, 1321–1342. [CrossRef]

29. Demirbaş, A. Biomass resource facilities and biomass conversion processing for fuels and chemicals. *Energy Convers. Manag.* **2001**, *42*, 1357–1378. [CrossRef]

30. Caputo, A.C.; Palumbo, M.; Pelagagge, P.M.; Scacchia, F. Economics of biomass energy utilization in combustion and gasification plants: Effects of logistic variables. *Biomass Bioenergy* **2005**, *28*, 35–51. [CrossRef]

31. Hall, D.O.; Scrase, J.I. Will biomass be the environmentally friendly fuel of the future? *Biomass Bioenergy* **1998**, *15*, 357–367. [CrossRef]

32. McKendry, P. Energy production from biomass (part 1): Overview of biomass. *Bioresour. Technol.* **2002**, *83*, 37–46. [CrossRef]

33. Shahrukh, H.; Oyedun, A.O.; Kumar, A.; Ghiasi, B.; Kumar, L.; Sokhansanj, S. Comparative net energy ratio analysis of pellet produced from steam pretreated biomass from agricultural residues and energy crops. *Biomass Bioenergy* **2016**, *90*, 50–59. [CrossRef]

34. Zhu, J.Y.; Zhuang, X.S. Conceptual net energy output for biofuel production from lignocellulosic biomass through biorefining. *Prog. Energy Combust. Sci.* **2012**, *38*, 583–598. [CrossRef]

35. Timmons, D.; Mejía, C.V. Biomass energy from wood chips: Diesel fuel dependence? *Biomass Bioenergy* **2010**, *34*, 1419–1425. [CrossRef]

36. Repo, A.; Tuovinen, J.P.; Liski, J. Can we produce carbon and climate neutral forest bioenergy? *GCB Bioenergy* **2015**, *7*, 253–262. [CrossRef]

37. Berndes, G.; Abts, B.; Asikainen, A.; Cowie, A.; Dale, V.; Egnell, G.; Lindner, M.; Marelli, L.; Paré, D.; Pingoud, K.; et al. *Forest Biomass, Carbon Neutrality and Climate Change Mitigation*; European Forest Institute: Joensuu, Finland, 2016.

38. Bright, R.M.; Cherubini, F.; Astrup, R.; Bird, N.; Cowie, A.L.; Ducey, M.J.; Marland, G.; Pingoud, K.; Savolainen, I.; Strømman, A.H. A comment to "Large-scale bioenergy from additional harvest of forest biomass is neither sustainable nor greenhouse gas neutral": Important insights beyond greenhouse gas accounting. *GCB Bioenergy* **2012**, *4*, 617–619. [CrossRef]

39. Cowie, A.; Berndes, G.; Jungigner, M.; Ximenes, F. *Response to Chatham House Report "Woody Biomass for Power and Heat: Impacts on the Global Climate"*; IEA Bioenergy: Lismore, Australia, 2017.

40. Helin, T.; Sokka, L.; Soimakallio, S.; Pingoud, K.; Pajula, T. Approaches for inclusion of forest carbon cycle in life cycle assessment—A review. *GCB Bioenergy* **2013**, *5*, 475–486. [CrossRef]

41. Pingoud, K.; Ekholm, T.; Soimakallio, S.; Helin, T. Carbon balance indicator for forest bioenergy scenarios. *GCB Bioenergy* **2016**, *8*, 171–182. [CrossRef]

42. Cherubini, F.; Peters, G.P.; Berntsen, T.; Strømman, A.H.; Hertwich, E. CO2 emissions from biomass combustion for bioenergy: Atmospheric decay and contribution to global warming. *GCB Bioenergy* **2011**, *3*, 413–426. [CrossRef]

43. Cherubini, F.; Bird, N.D.; Cowie, A.; Jungmeier, G.; Schlamadinger, B.; Woess-Gallasch, S. Energy- and greenhouse gas-based LCA of biofuel and bioenergy systems: Key issues, ranges and recommendations. *Resour. Conserv. Recycl.* **2009**, *53*, 434–447. [CrossRef]

44. IPCC. *IPCC Special Report on Renewable Energy Sources and Climate Change Mitigation*; IPCC: Cambridge, UK; New York, NY, USA, 2011.

45. Sochacki, S.J.; Harper, R.J.; Smettem, K.R.J. Bio-mitigation of carbon following afforestation of abandoned salinized farmland. *GCB Bioenergy* **2012**, *4*, 193–201. [CrossRef]

46. Berndes, G.; Hoogwijk, M.; Van Den Broek, R. The contribution of biomass in the future global energy supply: A review of 17 studies. *Biomass Bioenergy* **2003**, *25*, 1–28. [CrossRef]

47. Berndes, G.; Bird, N.; Cowie, A. *Bioenergy, Land Use Change and Climate Change Mitigation Background Technical Report*; IEA Bioenergy: Goteborg, Sweden, 2011.

48. Cummins, J.; Skennar, C.; Cassidy, M.; Palmer, G.; Capill, L. Using small hardwood logs to produce liquid fuels and electricity. *Aust. For.* **2016**, *79*, 189–195. [CrossRef]

49. Threlfall, C.G.; Law, B.S.; Peacock, R.J. Benchmarks and predictors of coarse woody debris in native forests of eastern Australia. *Austral Ecol.* **2019**, *44*, 138–150. [CrossRef]

50. Ximenes, F.; Stephens, M.; Brown, M.; Law, B.; Mylek, M.; Schirmer, J.; Sullivan, A.; McGuffog, T. Mechanical fuel load reduction in Australia: A potential tool for bushfire mitigation. *Aust. For.* **2017**, *80*, 1–11. [CrossRef]

51. Murphy, H.T.; O'Connell, D.A.; Raison, R.J.; Warden, A.C.; Booth, T.H.; Herr, A.; Braid, A.L.; Crawford, D.F.; Hayward, J.A.; Jovanovic, T.; et al. Biomass production for sustainable aviation fuels: A regional case study in Queensland. *Renew. Sustain. Energy Rev.* **2015**, *44*, 738–750. [CrossRef]

52. Hayward, J.A.; O'Connell, D.A.; Raison, R.J.; Warden, A.C.; O'Connor, M.H.; Murphy, H.T.; Booth, T.H.; Braid, A.L.; Crawford, D.F.; Herr, A.; et al. The economics of producing sustainable aviation fuel: A regional case study in Queensland, Australia. *GCB Bioenergy* **2015**, *7*, 497–511. [CrossRef]

53. Farine, D.R.; O'Connell, D.A.; Raison, R.J.; May, B.M.; O'Connor, M.H.; Crawford, D.F.; Herr, A.; Taylor, J.A.; Jovanovic, T.; Campbell, P.K.; et al. An assessment of biomass for bioelectricity and biofuel, and for greenhouse gas emission reduction in Australia. *GCB Bioenergy* **2012**, *4*, 148–175. [CrossRef]

54. Ritson, P.; Sochacki, S. Measurement and prediction of biomass and carbon content of Pinus pinaster trees in farm forestry plantations, south-western Australia. *For. Ecol. Manag.* **2003**, *175*, 103–117. [CrossRef]

55. Specht, A.; West, P.W.W. Estimation of biomass and sequestered carbon on farm forest plantations in northern New South Wales, Australia. *Biomass Bioenergy* **2003**, *25*, 363–379. [CrossRef]

56. Bi, H.; Long, Y.; Turner, J.; Lei, Y.; Snowdon, P.; Li, Y.; Harper, R.; Zerihun, A.; Ximenes, F. Additive prediction of aboveground biomass for Pinus radiata (D. Don) plantations. *For. Ecol. Manag.* **2010**, *259*, 2301–2314. [CrossRef]

57. Wang, X.; Bi, H.; Ximenes, F.; Ramos, J.; Li, Y. Product and residue biomass equations for individual trees in rotation age Pinus radiata stands under three thinning regimes in New South Wales, Australia. *Forests* **2017**, *8*, 439. [CrossRef]

58. Ngugi, M.R.; Neldner, V.J.; Ryan, S.; Lewis, T.; Li, J.; Norman, P.; Mogilski, M. Estimating potential harvestable biomass for bioenergy from sustainably managed private native forests in Southeast Queensland, Australia. *For. Ecosyst.* **2018**, *5*, 1–15. [CrossRef]

59. Mendham, D.S.; Ogden, G.N.; Short, T.; O'Connell, T.M.; Grove, T.S.; Rance, S.J. Repeated harvest residue removal reduces E. globulus productivity in the 3rd rotation in south-western Australia. *For. Ecol. Manag.* **2014**, *329*, 279–286. [CrossRef]

60. Garcia_Florez, L.; Vanclay, J.K.; Glencross, K.; Nichols, J.D. Developing biomass estimation models for above-ground compartments in Eucalyptus dunnii and Corymbia citriodora plantations. *Biomass Bioenergy* **2019**, *130*, 105353. [CrossRef]

61. Ximenes, F.A.; Gardner, W.D.; Richards, G.P. Total above-ground biomass and biomass in commercial logs following the harvest of spotted gum (Corymbia maculata) forests of SE NSW. *Aust. For.* **2006**, *69*, 213–222. [CrossRef]

62. Ximenes, F.A.; Gardner, W.D.; Kathuria, A. Proportion of above-ground biomass in commercial logs and residues following the harvest of five commercial forest species in Australia. *For. Ecol. Manag.* **2008**, *256*, 335–346. [CrossRef]

63. Ghaffariyan, M.R.; Acuna, M.; Wiedemann, J.; Mitchell, R. Productivity of the Bruks chipper when harvesting forest biomass in pine plantations. *CRC For.* **2011**, *1*, 5.

64. Walsh, D.; Strandgard, M. Productivity and cost of harvesting a stemwood biomass product from integrated cut-to-length harvest operations in Australian Pinus radiata plantations. *Biomass Bioenergy* **2014**, *66*, 93–102. [CrossRef]

65. Ghaffariyan, M.R.; Brown, M.; Acuna, M.; Sessions, J.; Kuehmaier, M.; Wiedemann, J. Biomass harvesting in Eucalyptus plantations in Western Australia. *South. For. A J. For. Sci.* **2011**, *73*, 149–154.

66. Ghaffariyan, M.R. Remaining slash in different harvesting operation sites in Australian plantations. *Silva Balc.* **2013**, *14*, 83–93.

67. Ghaffariyan, M.R.; Sessions, J.; Brown, M. Collecting harvesting residues in pine plantations using a mobile chipper in Victoria (Australia). *Silva Balc.* **2014**, *15*, 81–95.

68. Ghaffariyan, M.R.; Apolit, R. Harvest residues assessment in pine plantations harvested by whole tree and cut-to-length harvesting methods (a case study in Queensland, Australia). *Silva Balc.* **2015**, *16*, 113–122.

69. Ghaffariyan, M.R.; Spinelli, R.; Magagnotti, N.; Brown, M. Integrated harvesting for conventional log and energy wood assortments: A case study in a pine plantation in Western Australia. *South. For. A J. For. Sci.* **2015**, *77*, 249–254. [CrossRef]

70. Ghaffariyan, M.R.; Acuna, M.; Brown, M. Analysing the effect of five operational factors on forest residue supply chain costs: A case study in Western Australia. *Biomass Bioenergy* **2013**, *59*, 486–493. [CrossRef]

71. May, B.; England, J.R.; Raison, R.J.; Paul, K.I. Cradle-to-gate inventory of wood production from Australian softwood plantations and native hardwood forests: Embodied energy, water use and other inputs. *For. Ecol. Manag.* **2012**, *264*, 37–50. [CrossRef]

72. Crawford, D.F.; O'Connor, M.H.; Jovanovic, T.; Herr, A.; Raison, R.J.; O'Connell, D.A.; Baynes, T. A spatial assessment of potential biomass for bioenergy in Australia in 2010, and possible expansion by 2030 and 2050. *GCB Bioenergy* **2016**, *8*, 707–722. [CrossRef]

73. Meadows, J.; Coote, D.; Brown, M. The Potential Supply of Biomass for Energy from Hardwood Plantations in the Sunshine Coast Council Region of South-East Queensland, Australia. *Small Scale For.* **2014**, *13*, 461–481. [CrossRef]

74. Rothe, A.; Moroni, M.; Neyland, M.; Wilnhammer, M. Current and potential use of forest biomass for energy in Tasmania. *Biomass Bioenergy* **2015**, *80*, 162–172. [CrossRef]

75. Gupta, R.; Sharma, L.K. The process-based forest growth model 3-PG for use in forest management: A review. *Ecol. Modell.* **2019**, *397*, 55–73. [CrossRef]

76. Rodriguez, L.C.; May, B.; Herr, A.; O'Connell, D. Biomass assessment and small scale biomass fired electricity generation in the Green Triangle, Australia. *Biomass Bioenergy* **2011**, *35*, 2589–2599. [CrossRef]

77. Ximenes, F.A.; George, B.H.; Cowie, A.; Williams, J.; Kelly, G. Greenhouse gas balance of native forests in New South Wales, Australia. *Forests* **2012**, *3*, 653–683. [CrossRef]

78. Woo, H.; Acuna, M.; Moroni, M.; Taskhiri, M.S.; Turner, P. Optimizing the location of biomass energy facilities by integrating Multi-Criteria Analysis (MCA) and Geographical Information Systems (GIS). *Forests* **2018**, *9*, 585. [CrossRef]

79. Cowie, A.L.; Gardner, D.W.; David Gardner, W.; Gardner, D.W. Competition for the biomass resource: Greenhouse impacts and implications for renewable energy incentive schemes. *Biomass Bioenergy* **2007**, *31*, 601–607. [CrossRef]

80. Strandgard, M.; Turner, P.; Mirowski, L.; Acuna, M. Potential application of overseas forest biomass supply chain experience to reduce costs in emerging Australian forest biomass supply chains—A literature review. *Aust. For.* **2019**, *82*, 1–9. [CrossRef]

81. Fung, P.; Kirschbaum, M.U.F.; Raison, R.J.; Stucley, C. The potential for bioenergy production from Australian forests, its contribution to national greenhouse targets and recent developments in conversion processes. *Biomass Bioenergy* **2002**, *22*, 223–236. [CrossRef]

82. Moroni, M.T. Simple models of the role of forests and wood products in greenhouse gas mitigation. *Aust. For.* **2013**, *76*, 50–57. [CrossRef]

83. Tucker, S.N.; Tharumarajah, A.; May, B.; England, J.; Paul, K.; Hall, M.; Mitchell, P.; Rouwette, R.; Seo, S.; Syme, M. *Life Cycle Inventory of Australian Forestry and Wood Products*; Forest & Wood Products Australia: Melbourne, Australia, 2009; Available online: https://www.fwpa.com.au/images/marketaccess/PNA008-0708_Research_Report_LCI_Timber_0.pdf (accessed on 21 November 2019).

84. England, J.R.; May, B.; Raison, R.J.; Paul, K.I. Cradle-to-gate inventory of wood production from Australian softwood plantations and native hardwood forests: Carbon sequestration and greenhouse gas emissions. *For. Ecol. Manag.* **2013**, *302*, 295–307. [CrossRef]

85. Pang, X.; Trubins, R.; Lekavicius, V.; Galinis, A.; Mozgeris, G.; Kulbokas, G.; Mörtberg, U. Forest bioenergy feedstock in Lithuania – Renewable energy goals and the use of forest resources. *Energy Strategy Rev.* **2019**, *24*, 244–253. [CrossRef]

86. Mansuy, N.; Paré, D.; Thiffault, E.; Bernier, P.Y.; Cyr, G.; Manka, F.; Lafleur, B.; Guindon, L. Estimating the spatial distribution and locating hotspots of forest biomass from harvest residues and fire-damaged stands in Canada's managed forests. *Biomass Bioenergy* **2017**, *97*, 90–99. [CrossRef]

87. Sacchelli, S.; Fagarazzi, C.; Bernetti, I. Economic evaluation of forest biomass production in central Italy: A scenario assessment based on spatial analysis tool. *Biomass Bioenergy* **2013**, *53*, 1–10. [CrossRef]

88. Verkerk, P.J.; Fitzgerald, J.B.; Datta, P.; Dees, M.; Hengeveld, G.M.; Lindner, M.; Zudin, S. Spatial distribution of the potential forest biomass availability in europe. *For. Ecosyst.* **2019**, *6*, 1–11. [CrossRef]

89. Comber, A.; Dickie, J.; Jarvis, C.; Phillips, M.; Tansey, K. Locating bioenergy facilities using a modified GIS-based location-allocation-algorithm: Considering the spatial distribution of resource supply. *Appl. Energy* **2015**, *154*, 309–316. [CrossRef]

90. Fernandes, U.; Costa, M. Potential of biomass residues for energy production and utilization in a region of Portugal. *Biomass Bioenergy* **2010**, *34*, 661–666. [CrossRef]

91. Noon, C.E.; Daly, M.J. GIS-based biomass resource assessment with BRAVO. *Biomass Bioenergy* **1996**, *10*, 101–109. [CrossRef]

92. Castellano, P.J.; Volk, T.A.; Herrington, L.P. Estimates of technically available woody biomass feedstock from natural forests and willow biomass crops for two locations in New York State. *Biomass Bioenergy* **2009**, *33*, 393–406. [CrossRef]

93. Ranta, T. Logging residues from regeneration fellings for biofuel production-a GIS-based availability analysis in Finland. *Biomass Bioenergy* **2005**, *28*, 171–182. [CrossRef]

94. Vasco, H.; Costa, M. Quantification and use of forest biomass residues in Maputo province, Mozambique. *Biomass Bioenergy* **2009**, *33*, 1221–1228. [CrossRef]

95. Zhan, F.B.; Chen, X.; Noon, C.E.; Wu, G. A GIS-enabled comparison of fixed and discriminatory pricing strategies for potential switchgrass-to-ethanol conversion facilities in Alabama. *Biomass Bioenergy* **2005**, *28*, 295–306. [CrossRef]

96. Zhang, F.; Johnson, D.M.; Sutherland, J.W. A GIS-based method for identifying the optimal location for a facility to convert forest biomass to biofuel. *Biomass Bioenergy* **2011**, *35*, 3951–3961. [CrossRef]
97. Freppaz, D.; Minciardi, R.; Robba, M.; Rovatti, M.; Sacile, R.; Taramasso, A. Optimizing forest biomass exploitation for energy supply at a regional level. *Biomass Bioenergy* **2004**, *26*, 15–25. [CrossRef]
98. Hiloidhari, M.; Baruah, D.C.; Singh, A.; Kataki, S.; Medhi, K.; Kumari, S.; Ramachandra, T.V.; Jenkins, B.M.; Thakur, I.S. Emerging role of Geographical Information System (GIS), Life Cycle Assessment (LCA) and spatial LCA (GIS-LCA) in sustainable bioenergy planning. *Bioresour. Technol.* **2017**, *242*, 218–226. [CrossRef] [PubMed]
99. Zhang, F.; Johnson, D.; Johnson, M.; Watkins, D.; Froese, R.; Wang, J. Decision support system integrating GIS with simulation and optimisation for a biofuel supply chain. *Renew. Energy* **2016**, *85*, 740–748. [CrossRef]
100. Zhang, F.; Wang, J.; Liu, S.; Zhang, S.; Sutherland, J.W. Integrating GIS with optimization method for a biofuel feedstock supply chain. *Biomass Bioenergy* **2017**, *98*, 194–205. [CrossRef]

Article

Influence of Fertilization and Rootstocks in the Biomass Energy Characterization of *Prunus dulcis* (Miller)

Alba Mondragón-Valero [1], Borja Velázquez-Martí [2,*], Domingo M. Salazar [1] and Isabel López-Cortés [1]

[1] Departamento Producción Vegetal, Universitat Politènica de València, Camino de Vera s/n, 46022 Valencia, Spain; almonva@upv.es (A.M.-V.); dsalazar@upv.es (D.M.S.); islocor@upv.es (I.L.-C.)

[2] Departamento de Ingeniería Rural y Agroalimentaria, Universitat Politènica de València, Camino de Vera s/n, 46022 Valencia, Spain

*   Correspondence: borvemar@upv.es; Tel.: +96-387-72-90 (ext. 72900)

Received: 16 March 2018; Accepted: 3 May 2018; Published: 8 May 2018

Abstract: The importance of replacing fossil fuels with new energy routes such as the use of biomass leads to the improvement of sources such as agricultural and forest systems through adequate management techniques.The selection of the vegetal material and the management practices can influence the properties and quality of the obtained biofuel. The properties of the biomass obtained from pruning almond trees (*Prunus dulcis* (Mill)) have been analyzed in this study. Two varieties were tested, Marcona and Vayro, with three rootstocks, GF305, GF677 and GN Garnem, under different fertilization systems. The quality of the biofuel was evaluated with respect to the chemical composition and gross calorific value. We observed that the variables that mostly influenced the gross calorific value of the biomass were the variety, the rootstock and, primarily, the variety-rootstock interaction. Marcona presented better biomass properties than Vayro. Trees grafted on GF305 obtained a higher gross calorific value than the ones grafted on GF677 and GN Garnem. The percentage of nitrogen highly depended on the fertilization treatment applied, with saccharides and aminoacid fertilization accumulating a higher level of nitrogen than the humic and fluvic fertilization.

Keywords: biomass; variety and rootstock selection; almond tree; agricultural practices

---

1. Introduction

Energy efficiency is a basic pillar when it comes to achieving a certain degree of social and environmental sustainability, guaranteeing an adequate level of energy security [1]. In this sense, the European Union has set achieving greater efficiency in the development of renewal energies as a priority objective [2,3], committing to increase the exploitation of renewable energy by 20% [4].

Under this premise, the research of fuels as an alternative to crude oil and coal has led scientists to analyze materials that come from agriculture and forest environments [5–7]. Among the biofuels obtained from these systems, lignocellulosic materials such as wood chips and pellets take on an important role [8–10], as they can be used in combustion boilers to produce heat in both industrial—especially in oil mills—and domestic applications [11–13]. Improved cogeneration systems have also been developed to simultaneously produce heat and electricity [14,15]. Another very relevant application is the use of lignocellulosic materials in pyrolysis processes [16–18]. The primary biofuels used in small-scale combustion systems commonly have a woody origin, and as the demand for these biofuels grows, the pressure on forest exploitation will increase, with possible negative environmental consequences, which is why it is important to find new sources from which to obtain biofuel [4].

Ignorance regarding the energetic properties of the lignocellulosic materials from agricultural environments has led to numerous investigations determining the calorific value, ash, and elemental composition of the wood of different crops, such as the orange tree [19–21], the olive tree [8,20,21], the almond tree [19,20,22], the apple tree [23], the vine [8,19,20], and even herbaceous plant remains from greenhouses [24,25]. The heating or calorific value is one of the most important aspects related to the use of biomass, as it expresses the energy content of the biomass fuel and is a key parameter that has been widely used for the development of calorific power prediction models based on elemental, proximal and structural composition [26–30], although unfortunately the accuracy of the correlations based on those analyses are generally not very high [31].

In fact, just as the quality of the fruit varies according to the cultivation techniques used, the energy properties of the wood can be affected by the growing conditions and agricultural practices, such as the variety or rootstock selection or the fertilization.

Considering the lack of accurate biomass standardization, especially in relation to physicochemical, process and environmental parameters, the study and choice of raw materials for achieving better process efficiencies is not devoid of difficulties [32]. If the agricultural practices for obtaining materials influence their composition, obviously they will also do so in their energetic properties. Therefore, a proper characterization is required for the adequate use of the wastes obtained for biofuel uses.

The aim of this work was to demonstrate this hypothesis on the wood of almond tree (*Prunus dulcis* (Mill.)), which is one of the most important crops in both the Mediterranean region and also the Californian coast, which generates a huge amount of available biomass for energy uses [22], and has potential calorific powers superior to other common sources of biomass such as the almond shell, the olive pomace from oil mills, or the pressed grape waste from the wine industry [32]. Two varieties were tested, Marcona and Vayro, with two rootstocks and different fertilizations. The novelty offered by this study consists of the introduction of new parameters, such as rootstock selection and fertilization, in the study and classification of biomass products.

If the hypothesis is true, it would be necessary to consider those practices that improve the obtained products for food production, on the one hand, and on the other, the energetic material coming from the pruning waste. The use of both resources could significantly contribute to regional and national bio-economies in the future [9,14].

2. Materials and Methods

2.1. Field Study

The research area was located on the east coast of Spain, in the province of Valencia (latitude 39°28′50″ N, longitude 0°21′59″ W). The region is characterized by an average annual temperature of 18.3 °C. Maximum temperatures (30.2 °C) are observed in the month of August, and minimum temperatures (7.1 °C) in January. The average rainfall is around 475 mm, with a relative humidity of 65%; the driest period corresponds with the month of August, and the rainiest with October [33]. For study purposes, a total of 108 individuals of Prunus dulcis (Mill.).

(Mill.) were selected. The procedure of each trial consisted in the selection of a minimum of 36 individuals for the different combinations of factors that could influence the generated waste: variety, pattern and fertilization type. Table 1 shows the factors studied and their different levels.

Table 1. Factors and levels analyzed for the characterization of almond trees biomass.

Factor	Number of Levels	Levels
Variety	2	Marcona Vayro
Rootstock	3	GF 305 GF 677 GN Garnem
Fertilization	3	Biostimulant 1 Control Biostimulant 2

2.2. Vegetal Material and Treatments

For this research, we analyzed two almond tree varieties (both of them were characterized following the UPOV norm TG/56/4 Corr): Marcona and Vayro. Marcona is an autochthonous variety from the east of Spain that is known worldwide for the high quality of its fruits. It is a highly ramified, medium-high vigor tree with low cold requirements [34]. Vayro is a late-flowering variety obtained through crosses within the IRTA (Instituto de Investigación y Tecnología Agroalimentarias) genetic hybridization program. Trees of this variety show very strong vigor and medium branch density [35].

The behavior of 3 rootstocks widely used in almond cultivation was tested. GF 305 rootstock is a medium-vigor frank peach pattern that adapts adequately to irrigated conditions [36] and is tolerant to drought. GF 677 comes from the interbreeding of peach trees (*Prunus persica* L. Batsch) and almond trees (*Prunus dulcis* Miller) obtained in France by the INRA (Institut National de la Recherche Agronomique) [37]. It is a vigorous rootstock with good agronomic behavior in both dry and irrigated conditions. Finally, GN Garnem, obtained by the Agricultural Research Service of the Government of Aragón (CITA-DGA) as the result of crossing between *Prunus dulcis* (Mill.) (cv. Garrigues) and *Prunus persica* L. Batsch (cv. Nemared). It stands out for its strong vertical growth, inducing greater vigor than GF 677 [38].

Regarding fertilization, the samples were subjected to 3 types of treatment. Biostimulant 1, which is mainly composed of saccharides and amino acids; Biostimulant 2, rich in humic and fulvic extract, and a control group of test plants that were only irrigated with water.

2.3. Laboratory Analysis

2.3.1. Proximate Analysis

The evaluation of the drying process was carried out according to the norm ISO 18134-2:2017 [39]. The process took place in oven drying conditions under controlled temperature (105 ± 2) °C. In order to avoid the loss of volatile compounds, the drying process did not exceed 24 h. Ash content and volatile compounds were determined according to the norms ISO 18122:2015 [40] and ISO 18123:2015 [41], respectively.

2.3.2. Determination of Gross Calorific Value and Elemental Composition Analysis

Gross calorific values (CGV) of samples were analyzed using a LECO AC500 Automatic Calorimeter, based on norm ISO 18125:2017 [42]. Carbon (C), Hydrogen (H) and Nitrogen (N) determinations were carried out according to norm ISO 16948:2015 [43].

3. Results and Discussion

An overview of the chemical composition of almond tree biomass was conducted in order to describe the biomass potential of this species and how the selection of the vegetal material (variety and rootstock) and management practices can influence the properties of the obtained biofuel.

Table 2 presents the values of proximate analysis, elemental composition and gross calorific value of the studied samples. All variables, except the percentage of hydrogen, present standard kurtosis and standard skewness between −2 and +2, ANOVA analysis can be performed, as they all follow normal distributions. The average gross calorific value obtained in our analysis shows similar results to those obtained in almond hulls and shells by Nhuchhen [44] or by Fernández-González [19] or Yin [26] for biomass. The results are also close to those found for other *Prunus* species, such as *Prunus Avium* L. [45] or *Prunus armeniaca* L. [46]. As can be observed, the concentrations of C and H were 48.11% and 5.77%, respectively, which are the same as those observed by Jenkins [47] for almond residues. Percentages of C, H, N are in the range of published data for other biomass groups (C = 42–71%; H = 3–11%; N = 0.1–12%) [48]. Low concentration of N, as is the case, is a positive characteristic, as high N produces a negative impact on the environment due to nitrogen dioxide emissions [19]. The greatest disadvantage of biomass when it is used as fuel is its high moisture content, which is inversely correlated with its calorific value [49]. Almond trees presented an average moisture content (8.40%) much lower than the average for woody species (20%) [50], which therefore makes it convenient for energetic uses.

Table 2. Characterization of examined biomass.

	Average	Standard Deviation	Standard Skewness	Standard Kurtosis	Minimum	Maximum
Carbon (%)	48.41	2.22	−1.99	2.01	41.50	53.50
Hidrogen (%)	5.77	0.30	−4.81	4.96	4.83	6.60
Nitrogen (%)	0.97	0.27	1.62	−0.59	0.44	1.67
C/H	8.40	0.38	−1.53	0.70	7.29	9.26
GCV (joule/g)	18503.44	555.99	−6.73	7.99	17067.37	19628.39
Moisture content (%)	8.36	1.83	1.87	0.98	5.78	12.84
Ash (%)	3.41	0.80	2.05	−1.39	2.15	5.01
Volatile compounds (%)	82.89	5.48	−1.88	0.58	69.59	92.74

The dependence of the studied variables is shown in Table 3. It has been proved in other studies that moisture content and percentages of C and H have a high influence on gross calorific value [45,51]. However, in our study, we observed that the variables that mostly influence the gross calorific value are not the composition of C, H, N, but variety, rootstock and, above all, variety-rootstock interaction. We found that the variable that influenced the percentage of N the most was the treatment applied. This result is consistent with other research [52,53] that concludes that fertilizers and pesticide doses are highly important for some elements, such as the nitrogen content.

Given the obtained results, a variance analysis was carried out in order to compare the influence of the different factors analyzed in this study.

Table 3. Pearson correlation coefficients analysis.

	Variety	Rootstock	Interaction Variety/Rootstock	Treatment	C	H	N	C/H	GCV	Moisture Content (%)	Ash (%)	Volatile Compounds (%)
Variety		0	0.9869 *	0	−0.4053 *	−0.1831	0.027	−0.1983 *	−0.6465 *	−0.2160 *	0.2163 *	−0.1180
Rootstock	0		0.1612	0	−0.072	0.3213 *	−0.4055 *	−0.4346 *	−0.4769 *	−0.3373 *	−0.2213 *	0.6926 *
Interaction variety/rootstock	0.9869 *	0.1612		0	−0.4116 *	−0.1289	−0.0387	−0.2658 *	−0.7149 *	−0.2675 *	0.1778	−0.0049
Treatment	0	0	0		−0.1932	−0.3167 *	−0.4427 *	0.1606	0.0797	0.1503	0.2787 *	0.1605
C	−0.4053 *	−0.072	−0.4116 *	−0.1932 *		0.5593 *	0.1868 *	0.3581 *	0.4168 *	−0.0617	−0.0048	0.2528 *
H	−0.1831	0.3213 *	−0.1289	−0.3167 *	0.5593 *		−0.1143	−0.5722 *	0.0851	−0.2663 *	−0.1574	0.4891 *
N	0.0270	−0.4055 *	−0.0387	−0.4427 *	0.1868 *	−0.1143		0.3199 *	0.2552 *	0.2706 *	0.2334 *	−0.4033 *
C/H	−0.1983 *	−0.4346 *	−0.2658 *	0.1606	0.3581 *	−0.5722 *	0.3199 *		0.3118 *	0.2426 *	0.1688	−0.3013 *
GCV	−0.6465 *	−0.4769 *	−0.7149 *	0.0797	0.4168 *	0.0851	0.2552 *	0.3118 *		0.2781 *	0.2354 *	−0.0938
Moisture content (%)	−0.216 *	−0.3373 *	−0.2675 *	0.1503	−0.0617	−0.2663 *	0.2706 *	0.2426 *	0.2781 *		0.0584	−0.4015 *
Ash (%)	0.2163 *	−0.2213 *	0.1778	0.2787 *	−0.0048	−0.1574	0.2334 *	0.1688	0.2354 *	0.0584		0.2146 *
Volatile compounds (%)	−0.118	0.6926*	−0.0049	0.1605	0.2528 *	0.4891 *	−0.4033 *	−0.3013 *	−0.0938	−0.4015 *	0.2146 *	

* Pairs of variables with *P*-values lower than 0.1; C: carbon (%); H: hydrogen (%); N: nitrogen (%); GCV: gross calorific value $(J \cdot g^{-1})$.

3.1. The Variety Factor

Plants show a huge amount of similarities in properties between species, yet with significant specific variations such as the ones related to combustion properties. In this research, we focused on contrasting whether the differences were also notable between varieties within the same species. Marcona and Vayro varieties were statistically compared by means of variance analysis. The results obtained are shown on Figure 1. As we can see, the gross calorific value of Marcona (18858.70 J·g^{-1}) was statistically higher than the one obtained by Vayro (18148.20 J·g^{-1}). The same results were reached in the evaluation of the carbon and the hydrogen contents, while no differences were found for the nitrogen content. Marcona has been described in previous research as a variety with very high residue productivity (13.91 kg dry matter/tree) [20], this variety could be of interest in order to obtain not only a higher level of biomass in weight per tree but also a higher energetic potential because of its superior gross calorific value.

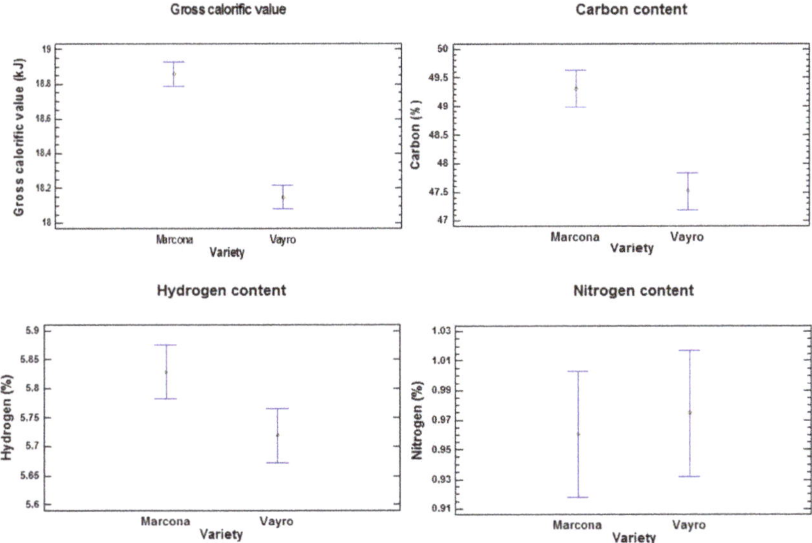

Figure 1. LSD intervals at 90% confidence level in the influence of the variety factor in residual biomass characteristics. (n = 108).

3.2. The Rootstock Factor

It has been widely studied that the selection of an appropriate rootstock is a key factor in fruit trees, as rootstocks can confer better adaptability to climatic and edaphic conditions and have influence on the plant mineral uptakes, the yield properties and efficiency, and the vigor of the grafted variety [54]. Figure 2 presents the statistical comparison of rootstocks by means of variance analysis. The rootstock with the highest gross calorific value was GF 305 (18850.70 J·g^{-1}) and the lowest gross calorific value was presented by rootstock GN Garnem (18148.20 J·g^{-1}). These differences between rootstocks cannot be attributed in our study to differences in carbon content, as we can see in Figure 2. This result would not correspond to those published by Demirbaş [55], in which the heating power was linked to the oxidation phase of the fuels, and in which carbon commonly dominated and overshadowed small changes of the hydrogen content. Although the contents of nitrogen in almond trees are low compared to other species [50], we can see that the rootstock factor affects nitrogen levels. In this study, the rootstock with the lowest nitrogen content is GN Garnem (0.84%) followed by GF 677 (0.95%). Nox and N$_2$O gases are generated when obtaining energy through biomass and are highly polluting [56]. These emissions directly depend on the nitrogen content of the biomass and must be as limited as possible.

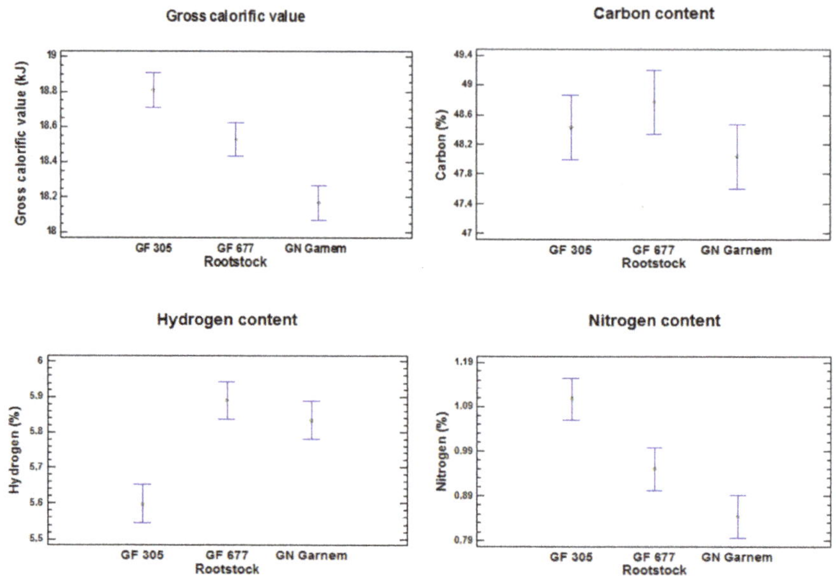

Figure 2. LSD intervals at 90% confidence level in the influence of the rootstock factor in residual biomass characteristics. ($n = 108$).

3.3. The Combination Variety-Rootstock Factor

Regardless of the rootstock on which they were grafted, Marcona samples presented greater values of gross calorific value than Vayro samples. The rootstock that induced the highest calorific value in both cultivars was GF 305. As seen in Table 4 and Figure 3, the gross calorific value of Marcona on GF 305 is superior to Marcona on GF 677 and Marcona on GN Garnem respectively. This same pattern of behavior is repeated in Vayro variety. The highest carbon contents were found in the combination variety-rootstock Marcona-GF 305 (50.45%) followed by Marcona-GF 677 (48.80%), Marcona-GN Garnem (48.68%) and Vayro-GF 677 (48.73%). As shown in Table 3, which contains the Pearson correlation coefficients, the carbon content is much more influenced by the grafted variety than by the rootstock. All variety-rootstock combinations presented similar values on the hydrogen content except for Vayro-GF 305, which was lower. This result coincides with the one obtained in the individualized study of the rootstocks (Figure 2). The lowest nitrogen content was identified in the Vayro-GN Garnem combination.

Table 4. Influence of the interaction variety-rootstock factor in residual biomass characteristics ($n = 108$; mean value $\pm$ standard deviation; mean values with different minor letters (a–e) differ significantly).

Interaction Variety-Rootstock	GCV (J)	Carbon (%)	Hydrogen (%)	Nitrogen (%)
Marcona 305	19270.66 ± 266.47 e	50.45 ± 1.77 c	5.77 ± 0.19 b	1.10 ± 0.14 d
Marcona 677	18807.45 ± 204.93 d	48.80 ± 1.07 b	5.87 ± 0.20 b	0.83 ± 0.30 ab
Marcona GN	18497.88 ± 375.76 c	48.68 ± 0.83 b	5.83 ± 0.24 b	0.94 ± 0.15 bc
Vayro 305	18349.76 ± 90.33 bc	46.41 ± 3.50 a	5.41 ± 0.39 a	1.11 ± 0.16 d
Vayro 677	18255.92 ± 238.11 b	48.73 ± 1.77 b	5.90 ± 0.14 b	1.06 ± 0.34 dc
Vayro GN	17839.02 ± 565.84 a	47.40 ± 0.80 a	5.83 ± 0.26 b	0.74 ± 0.19 a

In the same column, mean values with different minor letters differ significantly ($P < 0.1$).

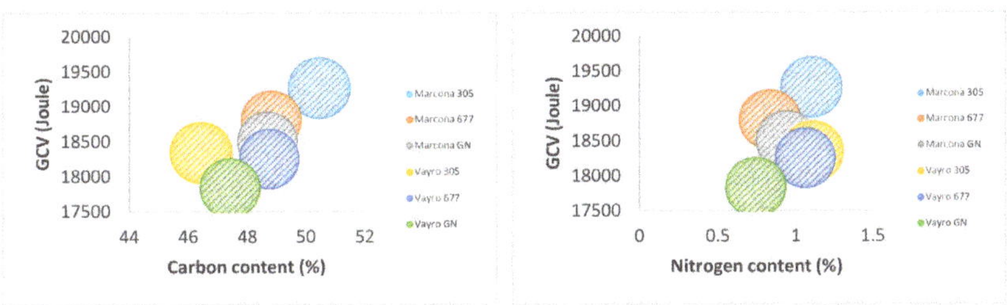

Figure 3. Dispersion diagram for gross calorific values (J·g^{-1}) according to the carbon and nitrogen content for each variety-rootstock combination.

3.4. The Treatment Factor

The direct uptake of organic compounds from the soil solution by the root system constitutes an important base of nutrients for the plant [57]. The purpose of this essay is to obtain greater knowledge on how the composition of fertilizers can modify the characteristics of residual biomass obtained in almond trees. No significative differences were found in terms of carbon content between the tested biostimulants and the control test; however, biostimulant 1 showed a higher carbon content than biostimulant 2. (Figure 4). The gross calorific value of the samples was not affected by fertilization, either. These results are in agreement with those obtained by Ercoli [58], in which the supply of N fertilizers in Miscanthus translated into a higher biomass yield, but did not affect the calorific value of the crop. Results obtained for biostimulant 1 regarding hydrogen and nitrogen contents are higher than those of biostimulant 2. These outcomes could be partially explained by the composition of the products, since biostimulant 1 mainly provides saccharides and amino acids (tryptophan, glutamic and GABA), which are very fast absorbing molecules [57]. Similar conclusions were reached by Mantineo [59], where extra doses of nitrogen fertilization led to a higher nitrogen concentration in the studied biomass. As high N produces a negative impact on the environment, due to nitrogen dioxide emissions [19], fertilizers that promote an increase of the percentage of N may be considered to be environmentally disadvantageous for biomasic uses.

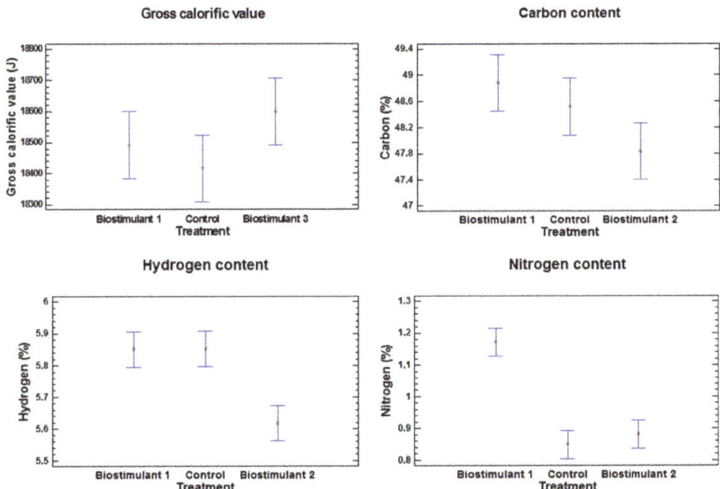

Figure 4. LSD intervals at 90% level of confidence in the influence of the treatment factor in residual biomass characteristics. (n = 108).

4. Conclusions

From a chemical point of view, almond tree pruning has suitable characteristics for use as a possible source, of biofuel as almond trees present high average gross calorific value with low moisture contents and low concentration of nitrogen.

The proposed hypothesis at the beginning of the trial is confirmed. In our study, the gross calorific value of the samples was more dependent on the varity and on the rootstock than on the C, H, N composition. It will therefore be advisable in the future to find those practices that maximize the yields of this crop both at the level of fruit production and at the level of biomass obtaining.

Among the studied varieties, Marcona presented higher energetical potential than Vayro. Important differences were found between the studied rootstocks. Although GF 305 presented the best results in terms of calorific power, it was also the rootstock with the highest nitrogen content. In contrast, GN had the lowest energy potential but also the lowest nitrogen content.

No significative differences were found in the gross calorific value of the samples according to the fertilization. However, percentage of nitrogen depended highly on the fertilization treatment applied, the saccharides and aminoacid fertilization accumulated a higher level of nitrogen than the humic and fluvic fertilization.

Author Contributions: B.V.-M., D.M.S. and I.L.-C. conceived and designed the experiments; A.M.-V. performed the experiments and analyzed the data; B.V.-M., D.M.S. and I.L.-C., A.M.-V. wrote the paper.

Acknowledgments: This work was funded by Project 20170734. Development of methods of quantification of riparian vegetation biomass for the management of channels of the Comunitat Valenciana. Dirección General de Universidades. Generalitat Valenciana (Spain).

Conflicts of Interest: The authors declare no conflict of interest.

References

1. Fernández-García, A.; Rojas, E.; Pérez, M.; Silva, R.; Hernández-Escobedo, Q.; Manzano-Agugliaro, F. A parabolic-trough collector for cleaner industrial process heat. *J. Clean. Prod.* **2015**, *89*, 272–285. [CrossRef]
2. Jacobsson, S.; Johnson, A. The diffusion of renewable energy technology: An analytical framework and key issues for research. *Energy Policy* **2000**, *28*, 625–640. [CrossRef]
3. Cruz-Peragon, F.; Palomar, J.M.; Casanova, P.J.; Dorado, M.P.; Manzano-Agugliaro, F. Characterization of solar flat plate collectors. *Renew. Sustain. Energy Rev.* **2012**, *16*, 1709–1720. [CrossRef]
4. Perea-Moreno, A.J.; Perea-Moreno, M.Á.; Hernandez-Escobedo, Q.; Manzano-Agugliaro, F. Towards forest sustainability in Mediterranean countries using biomass as fuel for heating. *J. Clean. Prod.* **2017**, *156*, 624–634. [CrossRef]
5. Esteban, L.S.; Carrasco, J.E. Biomass resources and costs: Assessment in different eu countries. *Biomass Bioenergy* **2011**, *35*, S21–S30. [CrossRef]
6. Sajdak, M.; Velázquez-Martí, B.; López-Cortés, I.; Fernández-Sarría, A.; Estornell, J. Prediction models for estimating pruned biomass obtained from *Platanus hispanica* Münchh. used for material surveys in urban forests. *Renew. Energy* **2014**, *66*, 178–184. [CrossRef]
7. Sajdak, M.; Velazquez-Marti, B. Estimation of pruned biomass form dendrometric parameters on urban forests: Case study of sophora japonica. *Renew. Energy* **2012**, *47*, 188–193. [CrossRef]
8. Velázquez-Martí, B.; Fernández-González, E.; López-Cortés, I.; Salazar-Hernández, D.M. Quantification of the residual biomass obtained from pruning of trees in Mediterranean olive groves. *Biomass Bioenergy* **2011**, *35*, 3208–3217. [CrossRef]
9. MacFarlane, D.W. Potential availability of urban wood biomass in Michigan: Implications for energy production, carbon sequestration and sustainable forest management in the U.S.A. *Biomass Bioenergy* **2009**, *33*, 628–634. [CrossRef]
10. Proskurina, S.; Junginger, M.; Heinimö, J.; Tekinel, B.; Vakkilainen, E. Global biomass trade for energy—Part 2: Production and trade streams of wood pellets, liquid biofuels, charcoal, industrial roundwood and emerging energy biomass. *Biofuels Bioprod. Biorefining* **2018**. [CrossRef]

11. Prando, D.; Renzi, M.; Gasparella, A.; Baratieri, M. Monitoring of the energy performance of a district heating CHP plant based on biomass boiler and ORC generator. *Appl. Therm. Eng.* **2015**, *79*, 98–107. [CrossRef]

12. Artese, C.; Schenone, G.; Bartolelli, V. Biomass Boilers for Household Heating. RES& RUE Dissemination Project. Available online: http://www.itabia.it/testi%20digitali/Dossier%20Caldaie%20a%20Biomassa.pdf (accessed on 12 October 2016). (In Spanish)

13. VYC. Industrial Biomass Boilers. Available online: http://vycindustrial.com/es/calderas/productos/calderas-industriales-de-biomasa/ (accessed on 16 October 2016). (In Spanish)

14. Uris, M.; Linares, J.I.; Arenas, E. Techno-economic feasibility assessment of a biomass cogeneration plant based on an Organic Rankine Cycle. *Renew. Energy* **2014**, *66*, 707–713. [CrossRef]

15. Uris, M.; Linares, J.I.; Arenas, E. Feasibility assessment of an Organic Rankine Cycle (ORC) cogeneration plant (CHP/CCHP) fueled by biomass for a district network in mainland Spain. *Energy* **2017**, *133*, 969–985. [CrossRef]

16. Haseli, Y.; van Oijen, J.A.; de Goey, L.P.H. Modeling biomass particle pyrolysis with temperature-dependent heat of reactions. *J. Anal. Appl. Pyrolysis* **2011**, *90*, 140–154. [CrossRef]

17. Morgan, H.M., Jr.; Bu, Q.; Liang, J.; Liu, Y.; Mao, H.; Shi, A.; Lei, H.; Ruan, R. A review of catalytic microwave pyrolysis of lignocellulosic biomass for value-added fuel and chemicals. *Bioresour. Technol.* **2017**, *230*, 112–121. [CrossRef] [PubMed]

18. Oh, W.D.; Lisak, G.; Webster, R.D.; Liang, Y.-N.; Veksha, A.; Giannis, A.; Moo, J.G.S.; Lim, Y.-W.; Lim, T.-T. Insights into the thermolytic transformation of lignocellulosic biomass waste to redox-active carbocatalyst: Durability of surface active sites. *Appl. Catal. B* **2018**, *233*, 120–129. [CrossRef]

19. Fernández-González, E. Análisis de los procesos de producción de biomasa residual procedente del cultivo de frutales mediterráneos. Cuantificación, cosecha y caracterización. Ph.D. Thesis, Universitat Politècnica de València, Valencia, Spain, 2010.

20. Velázquez-Martí, B.; López-Cortés, I.; Salazar-Hernández, D.; Callejón-Ferre, Á.J. Modeling the Calorific Value of Biomass from Fruit Trees Using Elemental Analysis Data. In *Biomass Volume Estimation and Valorization for Energy*; InTech: London, UK, 2017. [CrossRef]

21. Toklu, E. Biomass energy potential and utilization in Turkey. *Renew. Energy* **2017**, *107*, 235–244. [CrossRef]

22. Velázquez-Martí, B.; Fernández-González, E.; López-Cortés, I.; Salazar-Hernández, D.M. Quantification of the residual biomass obtained from pruning of trees in Mediterranean almond groves. *Renew. Energy* **2011**, *36*, 621–626. [CrossRef]

23. Winzer, F.; Kraska, T.; Elsenberger, C.; Kötter, T.; Pude, R. Biomass from fruit trees for combined energy and food production. *Biomass Bioenergy* **2017**, *107*, 279–286. [CrossRef]

24. Callejón-Ferre, A.J.; Carreño-Sánchez, J.; Suárez-Medina, F.J.; Pérez-Alonso, J.; Velázquez-Martí, B. Prediction models for higher heating value based on the structural analysis of the biomass of plant remains from the greenhouses of Almería (Spain). *Fuel* **2014**, *116*, 377–387. [CrossRef]

25. Barco, A.; Maucieri, C.; Borin, M. Root system characterization and water requirements of ten perennial herbaceous species for biomass production managed with high nitrogen and water inputs. *Agric. Water Manag.* **2018**, *196*, 37–47. [CrossRef]

26. Yin, C.-Y. Prediction of higher heating values of biomass from proximate and ultimate analyses. *Fuel* **2011**, *90*, 1128–1132. [CrossRef]

27. Vargas-Moreno, J.M.; Callejón-Ferre, A.J.; Pérez-Alonso, J.; Velázquez-Martí, B. A review of the mathematical models for predicting the heating value of biomass materials. *Renew. Sustain. Energy Rev.* **2012**, *16*, 3065–3083. [CrossRef]

28. Velázquez-Martí, B.; Sajdak, M.; López-Cortés, I.; Callejón-Ferre, A.J. Wood characterization for energy application proceeding from pruning *Morus alba* L., *Platanus hispanica* Münchh. and *Sophora japonica* L. in urban areas. *Renew. Energy* **2014**, *62*, 478–483. [CrossRef]

29. Bychkov, A.L.; Denkin, A.I.; Tikhova, V.D.; Lomovsky, O.I. Prediction of higher heating values of plant biomass from ultimate analysis data. *J. Therm. Anal. Calorim.* **2017**, *130*, 1399–1405. [CrossRef]

30. Osman, A.I.; Abdelkader, A.; Johnston, C.R.; Morgan, K.; Rooney, D.W. Thermal Investigation and Kinetic Modeling of Lignocellulosic Biomass Combustion for Energy Production and Other Applications. *Ind. Eng. Chem. Res.* **2017**, *56*, 12119–12130. [CrossRef]

31. Sheng, C.; Azevedo, J.L.T. Estimating the higher heating value of biomass fuels from basic analysis data. *Biomass Bioenergy* **2005**, *28*, 499–507. [CrossRef]
32. Álvarez, A.; Pizarro, C.; García, R.; Bueno, J.L. Spanish biofuels heating value estimation based on structural analysis. *Ind. Crop. Prod.* **2015**, *77*, 983–991. [CrossRef]
33. Agencia Estatal de Meteorología. Available online: http://www.aemet.es/es/serviciosclimaticos/datosclimatologicos (accessed on 20 October 2017).
34. Salazar, D.M.; Melgarejo-Moreno, P. *Cultivos leñosos: Frutales de zonas áridas. El cultivo del almendro*; Mundi-Prensa: Madrid, Spain, 2002.
35. Vargas, F.; Romero, M.; Clavé, J.; Vergés, J.; Santos, J.; Batlle, I. 'Vayro', 'Marinada', 'Constantí', and 'Tarraco' almonds. *HortScience* **2008**, *43*, 535–537.
36. Carrera, M.; Gomez-Aparasi, J. Rootstock influence on the performance of the peach variety 'Catherine'. In Proceedings of the IV International Peach Symposium, Bordeaux, France, 22–26 June 1997; pp. 573–578.
37. Bernhard, R.; Grasselly, C. Les pêchers x amandiers. *Arboric. Fruit* **1981**, *328*, 37–42.
38. Mondragón-Valero, A.; Lopéz-Cortés, I.; Salazar, D.M.; de Córdova, P.F. Physical mechanisms produced in the development of nursery almond trees (Prunus dulcis Miller) as a response to the plant adaptation to different substrates. *Rhizosphere* **2017**, *3*, 44–49. [CrossRef]
39. *ISO 18134-2:2017. Solid Biofuels–Determination of Moisture Content–Oven Dry Method–Part 2: Total Moisture–Simplified Method*; ISO: Geneva, Switzerland, 2017.
40. *ISO 18122:2015. Solid Biofuels—Determination of Ash Content*; ISO: Geneva, Switzerland, 2015.
41. *ISO 18123:2015. Solid Biofuels—Determination of the Content of Volatile Matter*; ISO: Geneva, Switzerland, 2015.
42. *ISO 18125:2017. Solid Biofuels. Determination of Calorific Value*; ISO: Geneva, Switzerland, 2017.
43. *ISO 16948:2015. Solid Biofuels. Determination of Total Content of Carbon, Hydrogen and Nitrogen*; Instrumental Methods; ISO: Geneva, Switzerland, 2015.
44. Nhuchhen, D.R.; Abdul Salam, P. Estimation of higher heating value of biomass from proximate analysis: A new approach. *Fuel* **2012**, *99*, 55–63. [CrossRef]
45. Telmo, C.; Lousada, J.; Moreira, N. Proximate analysis, backwards stepwise regression between gross calorific value, ultimate and chemical analysis of wood. *Bioresour. Technol.* **2010**, *101*, 3808–3815. [CrossRef] [PubMed]
46. Özçimen, D.; Ersoy-Meriçboyu, A. Characterization of biochar and bio-oil samples obtained from carbonization of various biomass materials. *Renew. Energy* **2010**, *35*, 1319–1324. [CrossRef]
47. Jenkins, B.; Baxter, L.L.; Miles, T.R., Jr.; Miles, T.R. Combustion properties of biomass. *Fuel Process. Technol.* **1998**, *54*, 17–46. [CrossRef]
48. Vassilev, S.V.; Baxter, D.; Andersen, L.K.; Vassileva, C.G. An overview of the chemical composition of biomass. *Fuel* **2010**, *89*, 913–933. [CrossRef]
49. Zhang, L.; Xu, C.C.; Champagne, P. Overview of recent advances in thermo-chemical conversion of biomass. *Energy Convers. Manag.* **2010**, *51*, 969–982. [CrossRef]
50. McKendry, P. Energy production from biomass (part 1): Overview of biomass. *Bioresour. Technol.* **2002**, *83*, 37–46. [CrossRef]
51. Callejón-Ferre, A.J.; Velázquez-Martí, B.; López-Martínez, J.A.; Manzano-Agugliaro, F. Greenhouse crop residues: Energy potential and models for the prediction of their higher heating value. *Renew. Sustain. Energy Rev.* **2011**, *15*, 948–955. [CrossRef]
52. Obernberger, I.; Biedermann, F.; Widmann, W.; Riedl, R. Concentrations of inorganic elements in biomass fuels and recovery in the different ash fractions. *Biomass Bioenergy* **1997**, *12*, 211–224. [CrossRef]
53. Nordin, A. Chemical elemental characteristics of biomass fuels. *Biomass Bioenergy* **1994**, *6*, 339–347. [CrossRef]
54. Ercisli, S.; Esitken, A.; Orhan, E.; Ozdemir, O. Rootstocks used for temperate fruit trees in Turkey: An overview. *Sodininkyste ir Darzininkyste* **2006**, *25*, 27–33.
55. Demirbaş, A. Relationships between lignin contents and heating values of biomass. *Energy Convers. Manag.* **2001**, *42*, 183–188. [CrossRef]
56. Kuhlbusch, T.A.; Lobert, J.M.; Crutzen, P.J.; Warneck, P. Molecular nitrogen emissions from denitrification during biomass burning. *Nature* **1991**, *351*, 135–137. [CrossRef]
57. Owen, A.G.; Jones, D.L. Competition for amino acids between wheat roots and rhizosphere microorganisms and the role of amino acids in plant N acquisition. *Soil Biol. Biochem.* **2001**, *33*, 651–657. [CrossRef]

58. Ercoli, L.; Mariotti, M.; Masoni, A.; Bonari, E. Effect of irrigation and nitrogen fertilization on biomass yield and efficiency of energy use in crop production of Miscanthus. *Field Crop. Res.* **1999**, *63*, 3–11. [CrossRef]
59. Mantineo, M.; D'Agosta, G.M.; Copani, V.; Patanè, C.; Cosentino, S.L. Biomass yield and energy balance of three perennial crops for energy use in the semi-arid Mediterranean environment. *Field Crop. Res.* **2009**, *114*, 204–213. [CrossRef]

Article

Measuring the Regional Availability of Forest Biomass for Biofuels and the Potential of GHG Reduction

Fengli Zhang [1], Dana M. Johnson [2], Jinjiang Wang [1,*], Shuhai Liu [1,*] and Shimin Zhang [1]

[1]   College of Mechanical and Transportation Engineering, China University of Petroleum,
    Beijing 102249, China; fengliz14@163.com (F.Z.); zsm7481976@163.com (S.Z.)
[2]   School of Business and Economics, Michigan Technological University, Houghton, MI 49931, USA;
    dana@mtu.edu
*   Correspondence: jwang@cup.edu.cn (J.W.); liu_shu_hai@163.com (S.L.);
    Tel.: +86-010-8973-3956 (J.W.); +86-010-8973-3647 (S.L.)

Received: 7 November 2017; Accepted: 8 January 2018; Published: 15 January 2018

Abstract: Forest biomass is an important resource for producing bioenergy and reducing greenhouse gas (GHG) emissions. The State of Michigan in the United States (U.S.) is one region recognized for its high potential of supplying forest biomass; however, the long-term availability of timber harvests and the associated harvest residues from this area has not been fully explored. In this study time trend analyses was employed for long term timber assessment and developed mathematical models for harvest residue estimation, as well as the implications of use for ethanol. The GHG savings potential of ethanol over gasoline was also modeled. The methods were applied in Michigan under scenarios of different harvest solutions, harvest types, transportation distances, conversion technologies, and higher heating values over a 50-year period. Our results indicate that the study region has the potential to supply 0.75–1.4 Megatonnes (Mt) dry timber annually and less than 0.05 Mt of dry residue produced from these harvests. This amount of forest biomass could generate 0.15–1.01 Mt of ethanol, which contains 0.68–17.32 GJ of energy. The substitution of ethanol for gasoline as transportation fuel has potential to reduce emissions by 0.043–1.09 Mt CO_{2eq} annually. The developed method is generalizable in other similar regions of different countries for bioenergy related analyses.

Keywords: timber; harvest residues; ethanol; GHG savings; Michigan

1. Introduction

In the United States (U.S.), transportation accounts for 69.8% of U.S. petroleum consumption [1] and 27% of total greenhouse gas (GHG) emissions in 2013. The transportation sector is the second largest contributor of U.S. GHG emissions following the electricity sector [2]. Increasing concerns that are associated with depletion of fossil fuels and global warming have imposed pressure on companies in the U.S. transportation sector and stimulated the evaluation of different alternative energy resources [3,4]. Bioethanol and biodiesel from lignocellulosic biomass (e.g., agricultural residues, woody biomass, and energy crops) could serve as partial replacement for petroleum based gasoline and diesel, respectively, and therefore, would help to minimise GHG emissions and achieve environmental goals.

There is a vast literature on biomass potential analysis worldwide. Hernandez et al. [5] assessed the theoretical and technical potential of available woody biomass for energy use with a regional case study in the north and central-south part of Mexico. Crawford et al. [6] conducted a spatial assessment of potentially available biomass for bioenergy in Australia in 2010, 2030, and 2050 for different types of biomass sources, including pulpwood and residues etc. Zhang et al. [7] explored the quantity and distribution of forest biomass in China based on forestry statistics data. Woch et al. [8] studied the potential of forest woody waste biomass for energy use in eastern Poland.

The State of Michigan (MI) offers significant potential for supplying forest-derived feedstocks [9] with annual growth far exceeding removals plus mortality in most timberland areas [9,10]. At the same time, a recent decline in traditional roundwood industries (e.g., pulp and paper, lumber) have revealed a new opportunity for the sustainable use of forest resources [11]. However, the estimates of feedstock availability suffer substantial uncertainties [12,13], such as the landowners' willingness and acceptance to harvest [14–16], the accessible with roads to harvest [17], the economic performance of feedstock supply chain (consisting of feedstock harvesting, transportation, and storage, etc.), as well as the delivered feedstock price [9,18]. All of these constraints should be considered when approximating long term biomass availability and estimating the corresponding biofuels potential.

Jakes and Smith [19] predicted Michigan's timber yields between 1980 and 2010. Sherrill and MacFarlane [20] assessed potential availability of urban forests, including wood residues and saw timber, for a 13-county region of Lower Michigan in 2007. MacFarlane [21] extended prior research in 2008 by examining potential urban tree biomass availability and the associated implications for energy production, carbon sequestration, and sustainable forest management. Mueller et al. [22] provided a snapshot of Michigan's woody biomass supply in 2010. Brunner et al. [23] assessed cellulosic ethanol production from the perspective of landscape scale net carbon, which is the tradeoff between the displacement of fossil fuel carbon emissions by biofuels and the high rates of carbon storage in aggrading forest stands. Gahagan et al. [24] evaluated carbon fluxes, storage, and harvest removals through 60 years of stand development in red pine plantations and mixed hardwood stands in Northern Michigan. Other studies revealed the availability of timber and residue in the Lake States region of Minnesota (MN), Wisconsin (WI), and Michigan (MI) [3,19]. Kukrety et al. [3] assessed sustainable forest biomass availability, likely harvest levels over a 100-year period, and bioenergy implications for the northern Lake States region. Becker et al. [19] examined current and projected resource needs for forest biomass in the Lake States.

There are also extensive scientific studies investigating GHG emissions mitigation potential of biomass resources worldwide. For example, Weldemichael and Assefaab [25] reported 11–15% of GHG emissions reduction by 2030, with the utilization of agricultural and forest biomass resources for energy production in Alberta. Veronika et al. [26] developed a GHG emission mitigation supply curves assuming a large-scale biomass use in Poland. Winchester and Reilly [27] assessed the contribution of biomass to emissions mitigation (16% less in basic policy case than the reference case) under a global climate policy.

In view of these studies, it is found that prior published research lacks a comprehensive study approximating long-term forest biomass availability and the associated uncertainties in the State of Michigan. In this study, the long-term (2015–2065) timber availability was derived from previous harvest trends for the Michigan. Harvest residues associated with growing stock volume cut or knocked down during harvest (including branches and tops) were approximated using the developed formulas considering a few of influential factors, including harvest types of all merchantable timber and the associated residue management options, and residue collection rate etc. Then, the potential forest biomass utilization for ethanol production and GHG reduction were also examined using the developed methods.

2. Materials and Methods

2.1. Study Area

The State of Michigan has a large biomass resource base that could be used as feedstock for biofuel facilities. In 2009, more than half (54%) of Michigan's land area was covered by forests [28]. The map of Michigan (Figure 1) shows the forest distribution on region basis in Michigan. A description of the study theme in this study is presented in Figure 2.

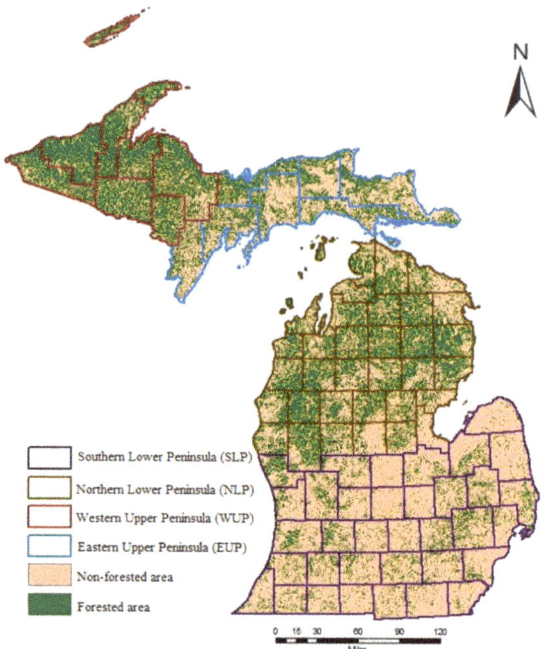

Figure 1. Michigan regions and forest (Based on study of [29]).

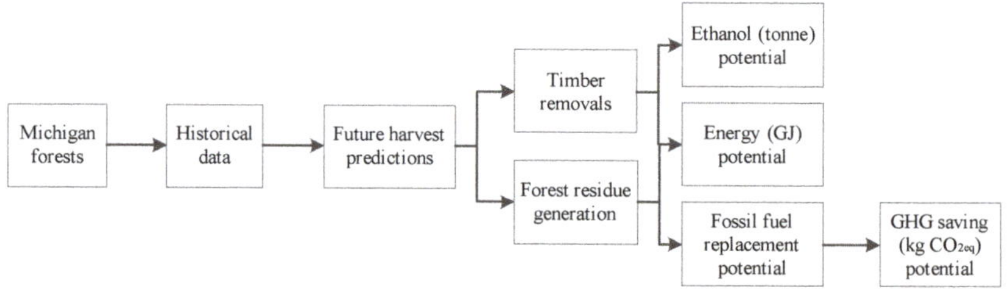

Figure 2. A description of the study theme.

2.2. Timber Volume Assessment and Prediction

Timber volume serves as a reasonable, though incomplete, measure of biomass [30]. Time trend analyses were employed by the U.S. Department of Agriculture (USDA) Forest Service timber assessments to understand how the timber supply today as it relates to the previous 50 years. The timber harvest trends were predicted by the functional form [31]:

$$\ln\left(HARVEST\right)_t = a_0 + a_1 TIME \tag{1}$$

where the dependent variable is the natural log of total harvest at time t, a_0 is a constant, a_1 is the slope of the trend line (rate of change of harvest divided by rate of change of time), and $TIME$ is the number of years since the start of the sample (current year less t_0 the date of the initial year) [31].

To use the above formula for long-term timber harvest estimates and forecasts, 26 years (1990–2015, fiscal year (1 October–30 September) annual timber sales data (in cords) were collected from literature. Cord is a volume unit commonly used in North-America. The Statewide Forest Resource Plan of 1983 promoted "stabilized timber supplies from public land", resulting in a stabilized forest products industry in Michigan [32]. This plan supports the suitability of using Equation (1) to estimate Michigan's timber harvest trend. As shown in Figure 3, a probability distribution describing the likely future timber harvests in Michigan by 2065 (over the 50-year time period) was created using the historic data collected. The calculated coefficients $a_0 = 13.428$ and $a_1 = 0.0077$. Thus, the Equation (1) can be rewritten as:

$$Q^T_{timber} = e^{13.428+0.0077TIME} \qquad (2)$$

where Q^T_{timber} (cords) is the annual timber harvest in year T and $TIME$ is the number of years since 1990.

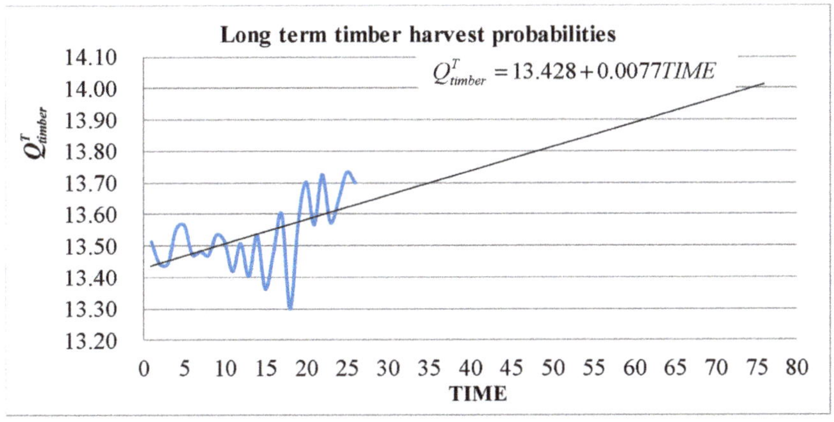

Figure 3. Graph showing the harvest trend developed from historic data for creating a probability distribution describing likely future harvests in Michigan.

2.3. Harvest Residues Estimation

Harvest residues are those associated with growing stock volume cut or knocked down during harvest (including branches and tops) [12]. In most cases, harvest residues are left at the harvest site, while only a limited quantity is collected from the landing point for energy purposes [33]. As a result, harvest residues are among the largest unused feedstock [19] with annual growth far exceeding removals [3], which presents an opportunity to increase harvest rates for strategic, economic, and forest health reasons [34]. The study [35] estimated logging residues by determining the proportional volume of tops and limbs in growing stock trees, which is approximately 17% of growing stock merchantable bole volume (tops/bole) for softwoods and 29% for hardwoods. In this study, the estimate method can be revised as the corresponding proportional volume of tops and limbs in timber harvest volume. Hardwoods and softwoods ratio was estimated to be 3:1 in Michigan based on the statistical information from [36].

Further, to estimate the quantity of logging residues for new biofuel facilities, it is critical to differentiate harvest types of all merchantable timber, since it is not a practical option by assuming clearcut treatment of all removals [37]. The main harvest types characterizing the logging industry in Michigan include [37,38]: (1) clearcutting all merchantable timber; and, (2) partial removal treatments (including 70% shelterwood and 30% selective cut). On the other hand, management of logging residue is also an important part of timber sale planning, which involves controlling the amount of residue remaining on the ground [39]. The residue management options (%) by harvest types were collected, as shown in Table 1. To explore use of residue in biofuel production and GHG reduction potential in

Michigan, a residue collection rate of 65% (onsite retention of 35%) was assumed for the removing residue management options in Table 1.

Table 1. Residue management options (%) by harvest types in Michigan (Based on study of [37]).

Residue Management Options	Symbol	Percent
Clearcut and leave residue	H_{cl}	27.8%
Clearcut and remove residue	H_{cr}	9.9%
Partial removal and leave residue	H_{pl}	50.9%
Partial removal and remove residue	H_{pr}	9.7%
Other method	H_o	1.7%
The sum		100%

Based on above assumptions, the quantity of logging residues can be estimated using the formula:

$$Q_{residue}^{T} = Q_{timber}^{T}(H_{cr} + H_{pr})(P_{timber}^{hard}p_{hard}^{residue} + P_{timber}^{soft}p_{soft}^{residue})\theta \tag{3}$$

where $Q_{residue}^{T}$ is the annual collectable residue volume (cords) in year T. P_{timber}^{hard} is the proportion of hardwoods in Michigan's timber harvest and P_{timber}^{soft} is the corresponding percent for softwoods. $p_{hard}^{residue}$ is the proportion of hardwoods in timber harvest, and $p_{soft}^{residue}$ is the corresponding percent for softwoods. θ is the residue collection rate.

For Michigan's case, the Equation (3) is simplified as:

$$Q_{residue}^{T,MI} = Q_{timber}^{T}(9.9\% + 9.7\%) \times (75\% \times 29\% + 25\% \times 17\%) \times 65\% = 0.033Q_{timber}^{T} \tag{4}$$

The Equation (4) shows that harvest residue collectable is only 3.3% of timber harvest. This is due to the low percent of removing residue options (9.9 + 9.7% = 19.6%) in Table 1. Most of the cases, logging residues are left onsite for retention.

2.4. Implications of Use for Ethanol Production and GHG Savings

2.4.1. Biomass Conversion Technologies

Different conversion technologies are available to convert forest and wood residues into a variety of biofuels and chemicals [40]. Conversion technologies of biomass to biofuels fall into two principal categories: thermo-chemical and bio-chemical technologies [41], which closely investigated several thermochemical and biological conversion processes employing three important metrics: mass ratio, energy ratio, and energy efficiency. The calculated metrics of ethanol from hardwood and softwood through different conversion processes are presented in Tables 2 and 3.

Table 2. Mass ratio of ethanol from hardwood and softwood through different conversion processes (Based on study of [41]).

Ethanol by Thermo-Chemical Process		
Hardwood	Softwood	Comments
0.49	0.52	Without catalytic methane reformation
0.73	0.76	With catalytic methane reformation
0.58	-	With feedstock used for process energy
Ethanol by Biochemical Process		
0.195	0.205	-

Table 3. Energy ratio of ethanol from hardwood and softwood through different conversion processes (Based on study of [41]).

Ethanol by Thermo-Chemical Process		
Hardwood	Softwood	Comments
0.58	0.6	Without catalytic methane reformation
0.86	0.88	With catalytic methane reformation
Ethanol by Biochemical Process		
0.23	0.215	-

Since there was no mass and energy ratio information provided for woody residue, the two metrics for the residues were assumed to be the same as timber. For Michigan's case, this assumption stands because residue volume accounts only 3.3% (obtained from Equation (4)) of timber volume.

2.4.2. Long Term Ethanol Probabilities

The estimates for the current and potential use of forest biomass for biofuel are based on volumes (cords). Since green weights are imprecise and highly variable, cubic foot volumes or dry weight are assumed to be more reliable estimates of inventory, growth, and removals and changes over time [11]. Therefore, a conversion to oven-dried tons (ODT) was calculated to convert the timber and residue volumes (cords) to dry weights. The conversion factor is 1.21 ODTs/cord, which is the average conversion factor that is used in the Forest Age Class Change Simulator (FACCS) simulation model [42]. Since 1 short ton = 0.907 metric tonne, the conversion factor is recalculated as $1.21 \times 0.907 = 1.097$ tonnes of dry matter per cord. By assuming that forest residues have the same conversion rate with timber roundwood, the mass of ethanol (M^T_{EtOH}, tonne) can be calculated as:

$$M^T_{EtOH} = 1.097(Q^T_{timber} + Q^T_{residue})(P^{hard}_{timber}\alpha^{mr}_{hard} + P^{soft}_{timber}\alpha^{mr}_{soft}) \tag{5}$$

where the α^{mr}_{hard} represents the mass ratio of ethanol from hardwood and α^{mr}_{soft} is the one from softwood.

With a density of 0.789 tonne/m^3 of ethanol [43], the tonnage of ethanol can be converted to liters through:

$$V^T_{EtOH} = 1,390(Q^T_{timber} + Q^T_{residue})(P^{hard}_{timber}\alpha^{mr}_{hard} + P^{soft}_{timber}\alpha^{mr}_{soft}) \tag{6}$$

where V^T_{EtOH} is the predicted volume (liters) of ethanol in year T.

For Michigan's case, Equation (4) and the values of P^{hard}_{timber} and P^{soft}_{timber} are substituted into Equations (5) and (6), the predicted mass (tonnes) and volume (liters) of ethanol in year T can be obtained by:

$$M^{T,MI}_{EtOH} = Q^T_{timber}(0.85\alpha^{mr}_{hard} + 0.28\alpha^{mr}_{soft}) \tag{7}$$

$$V^{T,MI}_{EtOH} = Q^T_{timber}(1,077\alpha^{mr}_{hard} + 355\alpha^{mr}_{soft}) \tag{8}$$

2.4.3. Long Term Energy Probabilities

Based on forest biomass predictions, the energy (E^T_{EtOH}, GJ) contained in the ethanol produced can be calculated as:

$$E^T_{EtOH} = 1.097(Q^T_{timber} + Q^T_{residue})(P^{hard}_{timber}\alpha^{mr}_{hard}\beta^{er}_{hard}HHV_{hard} + P^{soft}_{timber}\alpha^{mr}_{soft}\beta^{er}_{soft}HHV_{soft}) \tag{9}$$

where β^{er}_{hard} represent the energy ratio of ethanol from hardwood and β^{er}_{soft} is the one for softwood. HHV_{hard} (MJ/kg) is higher heat value for hardwood and HHV_{soft} (MJ/kg) is the one for softwood.

For Michigan's case, the long term energy probabilities can be calculated as:

$$E_{EtOH}^{T,MI} = Q_{timber}^{T}(0.85\alpha_{hard}^{mr}\beta_{hard}^{er}HHV_{hard} + 0.28\alpha_{soft}^{mr}\beta_{soft}^{er}HHV_{soft}) \tag{10}$$

2.4.4. GHG Savings Potential

For a more accurate estimation of GHG-related benefits of forest-based ethanol over gasoline, it is necessary to account for the emissions occur throughout the life cycle of ethanol from forest biomass pre-treatment and collection, transportation to biorefinery, conversion to ethanol, transportation to terminals, and the end use of ethanol. While the emissions from the end use or burning of ethanol are assumed to be carbon neutral because they are equivalent to the emissions that are captured during tree growth [44]. Therefore, the total annual GHG savings (GS^T, kg CO_{2eq}) can be expressed as the annual emission credits (G_{gas}^{T}, kg CO_{2eq}) obtained from replacing current gasoline minus the life cycle emissions (G_{gas}^{T}, kg CO_{2eq}) of ethanol in year T.

$$GS^T = G_{gas}^{T} - G_{EtOH}^{T} \tag{11}$$

The annual GHG emissions (G_{gas}^{T} and G_{EtOH}^{T}) is calculated by multiplying the amount of energy with an emission factor. The emission factor associated with combusting gasoline is 74.2 kg CO_{2eq}/GJ) [45]. The emission factor for ethanol takes the median life cycle emission per energy unit [25].

3. Results and Discussion

3.1. Potential of Biomass

The estimates for the current and potential use of forest biomass (timber, residues, and total in dry tonnes) for biofuel are illustrated in Figure 4. It is clear that timber is the major source of forest biomass and the availability of it increases steady in the foreseeable future. Harvest residues account for only a small proportion of the forest biomass and increase slowly over time. The results indicate that the study region has the potential to supply 0.75–1.4 Megatonnes (Mt) dry timber annually, and less than 0.05 Mt of dry residue produced from these harvests. These results are significantly smaller than that of [3], which did not consider the impacts of different harvest types on collecting residues.

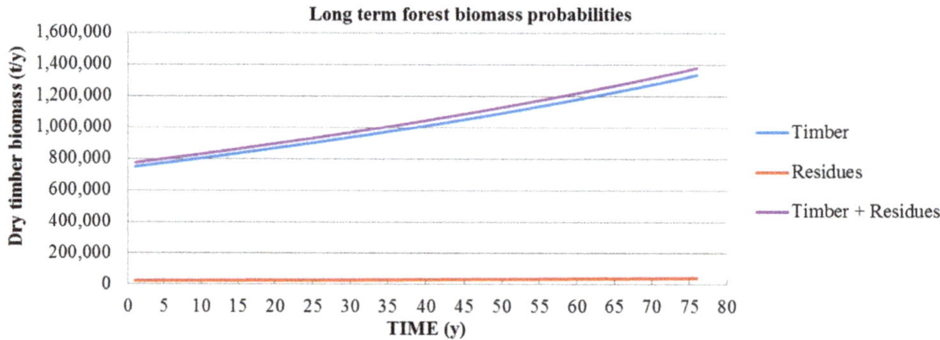

Figure 4. Long term forest biomass probabilities in Michigan.

3.2. Potential of Ethanol

Figure 5 presents the long term ethanol (tonne) probabilities in three scenarios: (1) ethanol by thermo-chemical process without catalytic methane reformation, (2) ethanol by thermo-chemical process with catalytic methane reformation, and (3) ethanol by biochemical process. As it can be seen in Figure 5, long term ethanol production shows steadily increasing trends for all three scenarios. As expected, the mass of ethanol in the scenario (1) achieves the highest when comparing with the other two scenarios due to no catalytic methane was reformed in the conversion process. The low mass ratio in scenario (2) resulted in low tonnage of ethanol production.

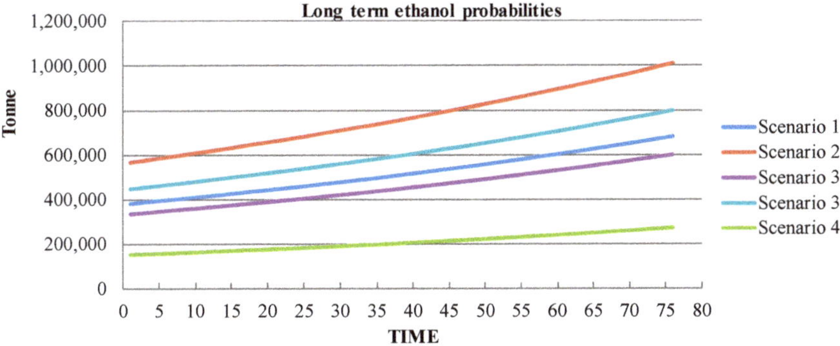

Figure 5. Long term ethanol probabilities in Michigan.

3.3. Potential of Energy and Emission Savings

According to [11], the commercial timber species within the Kinross supply region are categorized into five hardwood groups (Aspen, Maple, Oak, Upland Hardwoods, and Lowland Hardwoods) and three softwood groups (Pine, Upland Softwoods, and Lowland Softwoods). The Kinross supply region covers the Western Upper Peninsula (WUP, refer to Figure 1 for Michigan regions) in total, the Eastern Upper Peninsula (EUP) in part, and the Northern Lower Peninsula (NLP) in major part. As it can be seen from the Figure 1, the three regions (WUP, EUP, and NLP) are the main forest regions in Michigan. Thus, it is reasonable to take the timber species within the Kinross supply region as the timber species in Michigan. Since the HHV variation between species is usually smaller than the variations within one species [46], the average HHV (19.3 MJ/kg) collected for maple is used as the HHV for hardwood and the average HHV (20.9 MJ/kg) for pine as the HHV for softwood in this study. The results were illustrated in Figure 6. As expected, scenario 2 resulted in the most energy production due to the high mass and energy ratios.

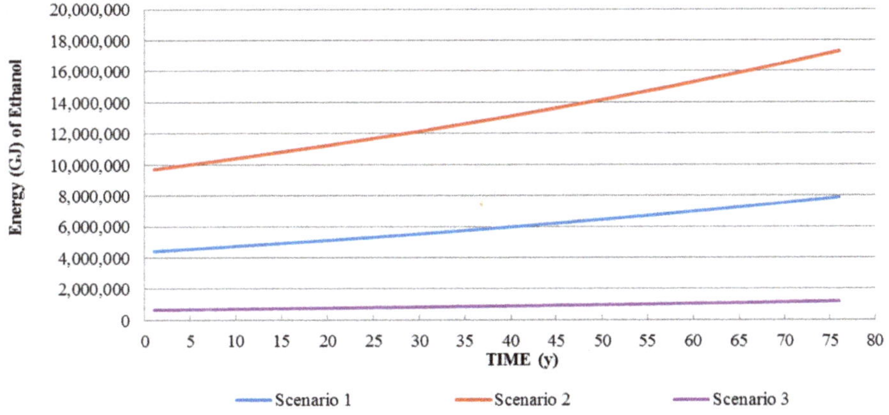

Figure 6. Long term energy probabilities of ethanol from forest biomass in Michigan.

Based on the long term energy probabilities of ethanol as shown in Figure 6, the total GHG savings are calculated and the predication results are shown in Figure 7.

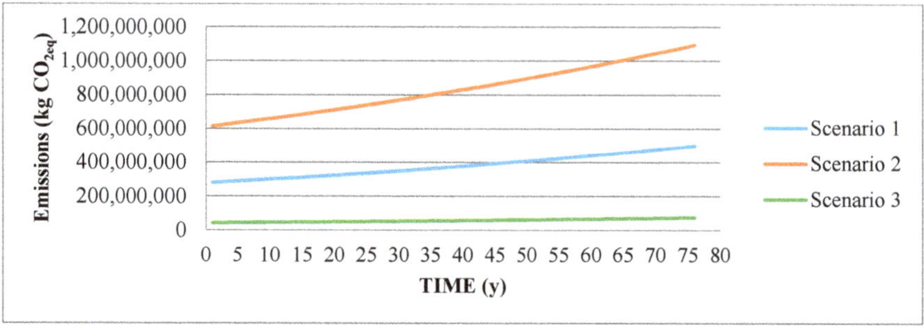

Figure 7. Long term emission savings probabilities of replacing gasoline with ethanol from forest biomass in Michigan.

4. Discussion

There is a vast of literature exploring the estimation methods of biomass potential. Since this study focused on timber and corresponding harvest residues instead of other types of biomass, only the literature studying forest biomass availability were examined. When compared to the volume estimation methods that are developed in this study, the 1998–1999 Silvicultural Analysis established an accounting framework and applied it to a 10-year (1999–2008) projection of the State Forest inventory in Michigan [32]. Kukrety et al. estimated the potential availability of roundwood and harvest residues using FIA inventory data and age-class progression techniques combined with average stand growth and yield models and applied it to the Lake States region [3]. To estimate potential annual yield from urban trees, it is a general method by evaluating the rate of urban trees becoming available for utilization. A vast of literature of potential wood biomass availability focused on growth rates of different vegetation type [47], while MacFarlane used mortality rate of urban trees and conducted a regional study of the potential availability of urban wood biomass in a 13-county area of Michigan [21]. Adams integrated published specific gravity and biomass distribution data with bolewood volumes predicted from regression equations relating these volumes to stand height and basal area to estimate forest biomass in northern lower Michigan [48].

As discussed above, the value for the prediction of biomass, ethanol, energy and emission savings vary depending up on regional parameters, scope of the study, assumptions made, and other factors that are considered for analysis [25]. This study predicted timber volume by collecting historical timber sales data from literature combing the time trend analyses from [31]. The energy potential was estimated based on heating value, which is a popular method for quality analysis of biomass [49]. This study conducted a comprehensive study approximating long-term forest biomass availability and associated energy production and emissions analysis to fill the gap by taking Michigan as one study region. The derived results are comparable to these studies discussed above.

5. Conclusions

Strong demand for oil in the United States (U.S.) for the transportation sector, multiple societal issues, including climate change concerns, heavy air pollution, high dependence on imports, and a multitude of security concerns have stimulated research in finding methods to reduce the use of fossil fuels for transportation fuel by substituting with renewable biomass. However, there is a lack of comprehensive regional estimates of available biomass, hindering the opportunity to maximise benefits. Using the State of Michigan in the United States as a case study, the long term harvest probabilities of timber were evaluated based on historic harvest data, as well as the implications of collecting residues and their use for ethanol production. The emission benefits of replacing fossil fuel with ethanol were also evaluated. Uncertainties that are considered include harvest solutions, harvest types, transportation distances of biomass/biofuel, biomass conversion technologies, and higher heating values, etc. The results indicate that the study region has the potential to supply 0.75–1.4 Mt dry timber annually and less than 0.05 Mt of dry residue produced from these harvests. This amount of forest biomass could generate 0.15–1.01 Mt of ethanol, which contains 0.68–17.32 GJ of energy. The substitution of ethanol for gasoline as transportation fuel has potential to reduce GHG emissions by 0.043–1.09 Mt CO_{2eq} annually. In addition to promoting energy security and reducing GHG emissions, the use of forest residues for energy would create additional income and employment opportunities in the forest based sector.

Acknowledgments: This research was supported by National Science Foundation of China (No. 71502173). This research acknowledges the financial support provided by National Key Research and Development Program of China (No. 2016YFC0802103), National Science Foundation of China (No. 51504274), and the Program for New Century Excellent Talents in University (No. NCET-13-1028). The Project-sponsored by SRF for ROCS, SEM.

Author Contributions: Fengli Zhang developed the research method, conducted the case analysis and wrote the paper. The rest of the authors helped collect the data and improve the wording of the paper.

Conflicts of Interest: The authors declare no conflict of interest.

References

1. Judith, G. Oil Sands up Close, 2012. Center for Climate and Energy Solutions (C2ES). Available online: http://www.c2es.org/energy/source/oil (accessed on 12 March 2016).
2. United States Environmental Protection Agency. (EPA). Sources of Greenhouse Gas Emissions. Available online: https://www3.epa.gov/climatechange/ghgemissions/sources/transportation.html (accessed on 15 March 2016).
3. Kukrety, S.; Wilson, D.C.; D'Amato, A.W.; Becker, D.R. Assessing sustainable forest biomass potential and bioenergy implications for the northern Lake States region, USA. *Biomass Bioenergy* **2015**, *81*, 167–176. [CrossRef]
4. Kannan, R.; Leong, K.C.; Osman, R.; Ho, H.K. Life cycle energy, emissions and cost inventory of power generation technologies in Singapore. *Renew. Sustain. Energy Rev.* **2007**, *11*, 702–715. [CrossRef]
5. Hernandez, U.F.; Jaeger, D.; Samperio, J.I. Bioenergy Potential and Utilization Costs for the Supply of Forest Woody Biomass for Energetic Use at a Regional Scale in Mexico. *Energies* **2017**, *10*, 1192. [CrossRef]
6. Crawford, D.F.; O'Connor, M.H.; Jovanovic, T.; Herr, A.; Raison, R.J.; O'Connell, D.A.; Baynes, T. A spatial assessment of potential biomass for bioenergy in Australia in 2010, and possible expansion by 2030 and 2050. *GCB Bioenergy* **2016**, *8*, 707–722. [CrossRef]
7. Zhang, C.X.; Zhang, L.M.; Xie, G.D. Forest Biomass Energy Resources in China: Quantity and Distribution. *Forests* **2015**, *6*, 3970–3984. [CrossRef]
8. Woch, F.; Hernik, J.; Wyrozumska, P.; Czesak, B. Residual Woody Waste Biomass as an Energy Source—Case Study. *Pol. J. Environ. Stud.* **2015**, *24*, 355–358. [CrossRef]
9. Department of Energy. *U.S. Billion-Ton Update: Biomass Supply for a Bioenergy and Bioproducts Industry*; Oak Ridge National Laboratory: Oak Ridge, TN, USA, 2011.
10. Van Deusen, P.C.; Roesch, F.A. Alternative definitions of growth and removals and implications for forest sustainability. *Forestry* **2008**, *81*, 176–182. [CrossRef]
11. Leefers, L.A.; Vasievich, M.J. Timber Resources and Factors Affecting Timber Availability and Sustainability for Kinross, Michigan 2011. Available online: http://www.michiganforestbiofuels.org/sites/default/files/Attachment%201_Feedstock%20Inventory%20report_v3_01072011.pdf (accessed on 20 August 2017).
12. Gan, J.; Smith, C.T. Availability of logging residues and potential for electricity production and carbon displacement in the USA. *Biomass Bioenergy* **2006**, *30*, 1011–1020. [CrossRef]
13. Galik, C.S.; Abt, R.C.; Wu, Y. Forest biomass supply in the Southeastern United States—Implications for industrial roundwood and bioenergy production. *J. For.* **2009**, *107*, 69–77.
14. Munsell, J.F.; Germain, R.H. Woody biomass energy: An opportunity for silviculture on nonindustrial private forestlands in New York. *J. For.* **2007**, *105*, 398–402.

15. Aguilar, F.; Garrett, H.E. Perspectives of Woody Biomass for Energy: Survey of State Foresters, State Energy Biomass Contacts, and National Council of Forestry Association Executives. *J. For.* **2009**, *107*, 297–306.

16. Becker, D.R.; Eryilmaz, D.; Klapperich, J.J.; Kilgore, M.A. Social availability of residual woody biomass from non-industrial private woodlands in Minnesota and Wisconsin. *Biomass Bioenergy* **2013**, *56*, 82–91. [CrossRef]

17. Butler, B.J.; Ma, Z.; Kittredge, D.B.; Catanzaro, P.F. Social versus biophysical availability of wood in the northern United States. *North. J. Appl. For.* **2010**, *27*, 151–159.

18. Becker, D.R.; Skog, K.; Hellman, A.; Halvorsen, K.E.; Mace, T. An outlook for sustainable forest bioenergy production in the Lake States. *Energy Policy* **2009**, *37*, 5687–5693. [CrossRef]

19. Jakes, P.J.; Smith, W.B. *Michigan's Predicted Timber Yields 1981–2010*; U.S. Department of Agriculture, Forest Service, North Central Forest Experiment Station: St. Paul, MN, USA, 1983; p. 98. Available online: Http://www.nrs.fs.fed.us/pubs/rp/rp_nc243.pdf (accessed on 3 May 2016).

20. Sherrill, S.B.; MacFarlane, D.W. Measures of Wood Resources in Lower Michigan: Wood Residues and the Saw Timber Content of Urban Forests. Available online: http://www.semircd.org/ash/research/sherrill_macfarlane_inventory_final.pdf (accessed on 3 May 2016).

21. MacFarlane, D.W. Potential availability of urban wood biomass in Michigan: Implications for energy production, carbon sequestration and sustainable forest management in the, U.S.A. *Biomass Bioenergy* **2009**, *33*, 628–634. [CrossRef]

22. Mueller, L.S.; Shivan, G.C.; Potter-Witter, K. *Michigan Woody Biomass Supply Snapshot*; Report to the Forestry Biofuels Statewide Collaboration Center; Michigan Economic Development Corporation: Lansing, MI, USA, 2010.

23. Brunner, A.; Currie, W.S.; Miller, S. Cellulosic ethanol production: Landscape scale net carbon strongly affected by forest decision making. *Biomass Bioenergy* **2015**, *83*, 32–41. [CrossRef]

24. Gahagan, A.; Giardina, C.P.; King, J.S.; Binkley, D.; Pregitzer, K.S.; Burton, A.J. Carbon fluxes, storage and harvest removals through 60 years of stand development in red pine plantations and mixed hardwood stands in Northern Michigan, USA. *For. Ecol. Manag.* **2015**, *337*, 88–97. [CrossRef]

25. Weldemichael, Y.; Assefa, G. Assessing the energy production and GHG (greenhouse gas) emissions mitigation potential of biomass resources for Alberta. *J. Clean. Prod.* **2016**, *112*, 4257–4264. [CrossRef]

26. Veronika, D.; van Dam, J.; Faaij, A. Estimating GHG emission mitigation supply curves of large-scale biomass use on a country level. *Biomass Bioenergy* **2007**, *31*, 46–65.

27. Winchester, N.; Reilly, J.M. The Contribution of Biomass to Emissions Mitigation under a Global Climate Policy. 2015. Available online: https://globalchange.mit.edu/sites/default/files/MITJPSPGC_Rpt273.pdf (accessed on 14 December 2017).

28. Forest Inventory and Analysis. FIA Standard Reports. 2009. Available online: http://fiatools.fs.fed.us/fido/standardrpt.html (accessed on 22 March 2016).

29. Mueller, L.S.; Potter-Witter, K. Regional variation of non-Industrial private forest owners in Michigan. In *Poster Abstract, Proceedings of the Society of American Foresters National Convention, Albuquerque, NM, USA, 27–30 October 2010*; Society of American Foresters: Bethesda, MD, USA, 2010; CD-ROM.

30. Michigan Forest Descriptors. Available online: http://mff.dsisd.net/TreeBasics/Descriptors.htm (accessed on 13 May 2016).

31. Darius, A.M.; Haynes, R.W.; Daigneault, A.J. Estimated Timber Harvest by U.S. Region and Ownership, 1950–2002. 2006. Available online: https://www.fs.usda.gov/treesearch/pubs/21682 (accessed on 9 January 2018).

32. Pedersen, L. Michigan State Forest Timber Harvest Trends. A Review of Recent Harvest Levels and Factors Influencing Future Levels. 2005. Available online: http://www.michigan.gov/documents/dnr/TimberHarvestTrends_173133_7.pdf (accessed on 10 May 2016).

33. D'Amato, A.W.; Bolton, N.W.; Blinn, C.R.; Ek, A.R. *Current Status and Long-Term Trends of Silvicultural Practices in Minnesota: A 2008 Assessment*; Staff Paper Series No. 205; Department of Forest Resources, University of Minnesota: St. Paul, MN, USA, December 2009; p. 58. Available online: http://mn.gov/frc/documents/council/UMN_silvics_Staffpaper205_2008report.pdf (accessed on 12 February 2016).

34. Sundstrom, S.; Nielsen-Pincus, M.; Moseley, C.; McCaffery, S. Woody biomass use trends, barriers, and strategies: Perspectives of US Forest Service Managers. *J. For.* **2012**, *110*, 16–24. [CrossRef]

35. Tessa Systems, LLC. Forest-Based Woody Biomass Assessment for Michigan's Upper Peninsula. Final Report Prepared for the Michigan Economic Development Corporation. Available online: http://mff.dsisd.net/biomass/BiomassDocs/Tessa-UP-2010.pdf (accessed on 12 February 2016).

36. Potter-Witter, K. Michigan Timber Available to Harvest. Available online: http://fbis.mtu.edu/caveats.pdf (accessed on 12 February 2016).
37. Abbas, D.; Handler, R.; Lautala, P.; Hartsough, B.; Dykstra, D.; Hembroff, L. A Survey Analysis of Forest Harvesting and Transportation Operations in Michigan. *Croat. J. For. Eng.* **2014**, *35*, 179–192.
38. Handler, R.M.; Shonnard, D.R.; Lautala, P.; Abbas, D.; Srivastava, A. Environmental impacts of roundwood supply chain options in Michigan: Life-cycle assessment of harvest and transport stages. *J. Clean. Prod.* **2014**, *76*, 64–73. [CrossRef]
39. Thomas, A.C. *Managing Logging Residue under the Timber Sale Contract*; Res. Note PNW-RN-348; U.S. Department of Agriculture, Forest Service, Pacific Northwest Forest and Range Experiment Station: Portland, OR, USA, 1980.
40. Srirangan, K.; Akawi, L.; Moo-Young, M.; Chou, C.P. Towards sustainable production of clean energy carriers from biomass resources. *Appl. Energy* **2012**, *100*, 172–186. [CrossRef]
41. Mcgill University & IATA. The 2nd Generation Biomass Conversion Efficiency. Available online: http://www.academia.edu/3166371/2nd_Generation_Biomass_Conversion_Efficiency (accessed on 12 May 2016).
42. Wilson, D.C.; Domke, G.M.; Ek, A.R. Forest Age Class Change Simulator (FACCS): A Spreadsheet-Based Model for Estimation of Forest Change and Biomass Availability. 2014. Available online: http://conservancy.umn.edu/bitstream/handle/11299/170674/Staffpaper228.pdf?sequence=3&isAllowed=y (accessed on 24 February 2016).
43. Murphy, J.D.; Mccarthy, K. Ethanol production from energy crops and wastes for use as a transport fuel in Ireland. *Appl. Energy* **2005**, *82*, 148–166. [CrossRef]
44. Cambero, C.; Sowlatia, T.; Pavel, M. Economic and life cycle environmental optimization of forest-based biorefinery supply chains for bioenergy and biofuel production. *Chem. Eng. Res. Des.* **2016**, *107*, 218–235. [CrossRef]
45. *DEFRA: 2008 Guidelines to Defra's GHG Conversion Factors Annex 1*; HMSO (Department for Environment Food and Rural Affairs): London, UK, 2007; Available online: http://www.aef.org.uk/downloads/ghg-cf-guidelines-annexes2008.pdf (accessed on 25 August 2017).
46. Penn State's Renewable and Alternative Energy Program. Renewable and Alternative Energy Fact Sheet: Characteristics of Biomass as a Heating Fuel. Available online: http://www.biomassinnovation.ca/pdf/factsheet_PennState_BiomassProperties.pdf (accessed on 18 December 2017).
47. Mead, D.J. Forest for energy and the role of planted trees. *Crit. Rev. Plant Sci.* **2005**, *24*, 407–421. [CrossRef]
48. Adams, P.W. Estimating biomass in northern lower Michigan forest stands. *For. Ecol. Manag.* **1982**, *4*, 275–286. [CrossRef]
49. Hossen, M.M.; Rahman, A.H.; Kabir, A.S.; Hasan, M.M.; Ahmed, S. Systematic assessment of the availability and utilization potential of biomass in Bangladesh. *Renew. Sustain. Energy Rev.* **2017**, *67*, 94–105. [CrossRef]

 *energies*

Article

Pelleting of Pine and Switchgrass Blends: Effect of Process Variables and Blend Ratio on the Pellet Quality and Energy Consumption

Jaya Shankar Tumuluru

Idaho National Laboratory, Energy Systems and Technologies Directorate, Bioenergy Technologies Department, 750 MK Simpson Blvd., Idaho Falls, ID 83415-3570, USA; JayaShankar.Tumuluru@inl.gov; Tel.: +1-208-526-0529

Received: 20 February 2019; Accepted: 18 March 2019; Published: 28 March 2019

Abstract: The blending of woody and herbaceous biomass can influence pellet quality and the energy consumption of the process. This work aims to understand the pelleting characteristics of 2-inch top-pine residue blended with switchgrass at high moisture content. The process variables tested are blend moisture content, length-to-diameter (L/D) ratio in the pellet die, and the blend ratio. A flat die pellet mill was also used in this study. The pine and switchgrass blend ratios that were tested include: (1) 25% 2-inch top pine residue with 75% switchgrass; (2) 50% 2-inch top pine residue with 50% switchgrass; and (3) 75% 2-inch top pine residue with 25% switchgrass. The pelleting process conditions tested included the L/D ratio in the pellet die (i.e., 1.5 to 2.6) and the blend moisture content (20 to 30%, w.b.). Analysis of experimental data indicated that blending 25% switchgrass with 75% 2-inch top pine residue and 50% switchgrass with 50% 2-inch top pine residue resulted in pellets with a bulk density of > 550 kg/m^3 and durability of > 95%. Optimization of the response surface models developed for process conditions in terms of product properties indicated that a higher L/D ratio of 2.6 and a lower blend-moisture content of 20% (w.b.) maximized bulk density and durability. Higher pine in the blends improved the pellet durability and reduced energy consumption.

Keywords: 2-inch top pine residue + switchgrass blends; pelleting process variables; pellet quality; specific energy consumption; response surface models; hybrid genetic algorithm

---

1. Introduction

According to the United States (U.S.) Department of Energy–Office of Energy Efficiency and Renewable Energy (DOE–EERE), agricultural crop and forest residues, grasses, and woody energy crops that are grown specifically for their energy, together with algae, industrial wastes, sorted municipal solid waste, urban wood waste, and food waste, are considered to be biomass, which can be used for the production of fuels and chemicals [1]. Liquid fuels produced from biomass can also act as a supplement to petroleum-based liquid transportation fuels, such as gasoline and diesel [1]. In addition, biomass can be used to produce valuable chemicals and electricity. According to the DOE Billion-Ton report [2], there are more than a billion tons of biomass produced annually in the U.S., which could sustainably be accessed for continuous biofuels production. A 2005 study conducted by both the U.S. Department of Agriculture (USDA)/DOE indicated that both woody and herbaceous perennial bioenergy crops should be considered for bioenergy and bioproducts production [3].

The southeastern United States is projected to supply almost 50% of the 16 billion gallons of advanced cellulosic biofuels mandated by the Renewable Fuels Standard [4]. Assuming typical yields of 80 gallons of fuel per dry ton and 4–8 dry tons/acre, this demand translates to roughly 200 million tons of biomass produced on 25 million acres [5]. This target is considered to be readily achievable, playing to the region's strength of lignocellulosic biomass production [5]. An analysis conducted by the DOE Logistics for Enhanced Attributes Feedstock (LEAF) project estimates that by using the Billion-Ton report as a base-case scenario, primary sources of available biomass are residues generated from forest-products industry and agricultural residues. However, in future, a diverse portfolio of forest, agriculture biomass and second generation bioenergy crops such as miscanthus, switchgrass, and energy cane are going to play a major role in biofuels production. One of the major cost contributors in bioenergy production is feedstock. To make bioenergy a reality, reducing feedstock cost and improving feedstock quality are important for commercial-scale implementation of biomass conversion technologies for fuels and chemical production.

1.1. Biomass Variability

Due to the diversity in the supply of available biomass, biorefineries are expected to receive varied feedstocks with different physical properties and chemical compositions. Biomass obtained from different regions, weather patterns, harvesting, handling and storage conditions, and crop varieties increases this variability factor even further. To address these variability issues, biorefineries need either to re-engineer their processes for each feedstock or to design systems with extreme tolerance, which can have a significant impact on the overall cost. One of the major limitations in using biomass at large scale is variability in the physical and chemical properties of the biomass and the seasonal and geographic availability of the biomass [6]. The variability in biomass physical and chemical properties limit its commercial-scale applications [7]. For example, biomass moisture variability influences grinder throughput and particle size distribution, which in turn causes inconsistent mass and heat transfer in conversion. Particle size variability creates feed-handling and conversion issues. For example, larger particle sizes (such as chips and coarsely ground herbaceous biomass) plug bins and augers and do not fully cook in digesters, thereby plugging downstream equipment. Fine particles influence ash composition, thus causing fire, explosion, and health hazards, plugging of weep holes in digesters, and creating inconsistent mass and heat transfer during biochemical and thermochemical conversion.

1.2. Blending

Blending is a common method that mixes different types of biomass to improve their physical properties and chemical composition. In their studies on the pelleting of woody and herbaceous biomass blended feedstocks, Yancey et al. [8] indicated that blending helps to reduce physical-property and chemical-composition variability in various biomass sources while producing a consistent feedstock. For example, different grades of coals are blended to reduce their sulfur and nitrogen content. Various high-ash biomass sources are blended with low-ash biomass sources for biopower generation. In the agricultural industry, grains are blended to adjust their moisture content. In the feed industry, ingredients are blended to maintain the nutrient content of the feed [9]. Ray et al. [10] suggested that biomass blending helps to overcome cost and quality limitations of biomass for biofuels production, while Edmunds et al. [11] suggested that the blending of different biomass sources helps to improve feedstock specifications. For thermochemical conversion, attributes of interest include carbon content, total ash, and specific minerals, density, and moisture content.

According to Ray et al. [10], Kenney et al. [12], and Thompson et al. [13], biomass blending helps to overcome challenges associated with feedstock quality, variability, supply, and cost. The major advantages of biomass blending are: (1) an increase in potential biomass supply for a given biorefinery area; (2) feedstock cost reduction; and (3) improvement in biomass flow and pelleting characteristics [8,14]. Recently, the blending of different sources of lignocellulosic biomass to produce feedstocks for thermochemical conversion has gained importance. For example, Mahadevan et al. [15] reported that blending switchgrass and southern pinewood resulted in bio-oils with low acidity and viscosity, but higher water content. The major challenges of blending biomass from various sources are these variabilities in biomass physical properties in terms of particle size, moisture, and density. These feedstock variability parameters can result in issues related to feeding, handling, transport, and storage.

According to Tumuluru [6], biomass pretreatment and preprocessing can help to overcome variability issues. Mechanical (e.g., size reduction, densification), chemical (e.g., ammonia fiber expansion, acid, alkali, ionic), and thermal (e.g., torrefaction, hydrothermal liquefaction) preprocessing and pretreatment help to address biomass variability in physical properties and chemical composition. In addition, Tumuluru [6,7], Tumuluru et al. [16,17], and Tumuluru and Yancey [18] suggested that mechanical preprocessing and thermal pretreatment of biomass helps to improve biomass physical properties (such as particle size distribution and bulk density), chemical properties (such as proximate and ultimate composition), and energy property (such as calorific value).

1.3. Densification

The low bulk density of biomass, which is typically in the range of 150–200 kg/m^3 for woody biomass [19] and 80–100 kg/m^3 for herbaceous biomass [16], limits its application at the commercial scale. The low bulk densities of biomass make biomass material difficult to store, transport, and interface with biorefinery infeed systems [20]. In general, high-bulk and low-energy-density biomass results in difficulty in feeding the biomass and reduces conversion efficiencies. Densification of biomass helps to overcome this limitation. According to Tumuluru et al. [16], the densification process is critical for producing a feedstock material suitable as a commodity product. Densification helps to overcome the physical properties variability issues, such as moisture, particle size distribution, and density. Densified biomass has improved handling and conveyance efficiencies throughout the supply system and biorefinery infeed, and improved feedstock uniformity and density.

Common biomass densification systems have been adapted from other highly efficient processing industries like feed, food, and pharmacy, and include: (1) a pellet mill; (2) a cuber; (3) a briquette press; (4) a screw extruder; (5) a tabletizer; and (6) an agglomerator [16,20]. The major challenge in biomass densification using the pelleting process is drying of the biomass to about 10–12% (w.b.) moisture content using conventional drying systems, such as a rotary dryer [21–23]. In their study on the validation of advanced feedstock supply systems, Searcy et al. [24] indicated that one of the major limitations biorefineries face in using high-moisture woody and herbaceous biomass for biofuels production is high preprocessing (i.e., for size reduction, drying, and densification) costs. Figure 1 indicates the different unit operations in a conventional pellet production process [25].

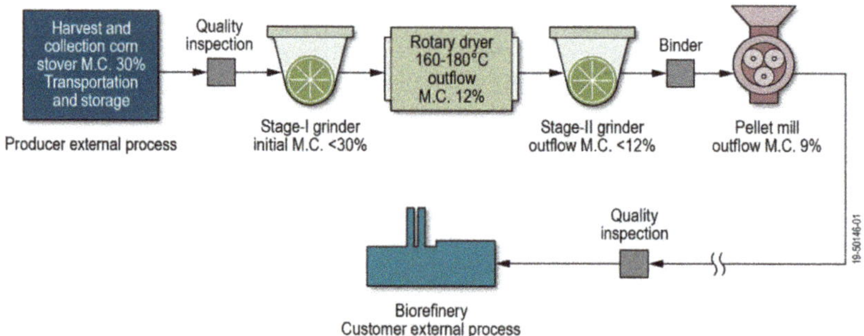

Figure 1. Various unit operation in the conventional pelleting process [25].

Techno-economic analysis indicated efficient moisture management is critical for reducing the preprocessing costs of biomass [26]. According to Pirraglia et al. [27], Sakkampang and Wongwuttanasatian [28], and Yancey et al. [8], the drying of biomass using rotary dryers is a significant energy-consuming unit operation in the pelleting process. According to Tumuluru [23], drying biomass from 10–30% (w.b.) for pelleting takes about 65% energy, whereas pelleting itself only requires about 8–9%, as shown in Figure 2 [23]. Another major limitation with high-temperature biomass drying for biofuels production is the emission of volatile organic emissions.

According to Granström [29] and Johansson and Rasmuson [30], woody biomass drying using a high-temperature rotary dryer results in the emission of volatiles and extractives that are suitable neither for human health nor the environment. When released into the environment, these emissions form photo-oxidants, which are dangerous if inhaled by humans and can also damage forests and the plant canopy.

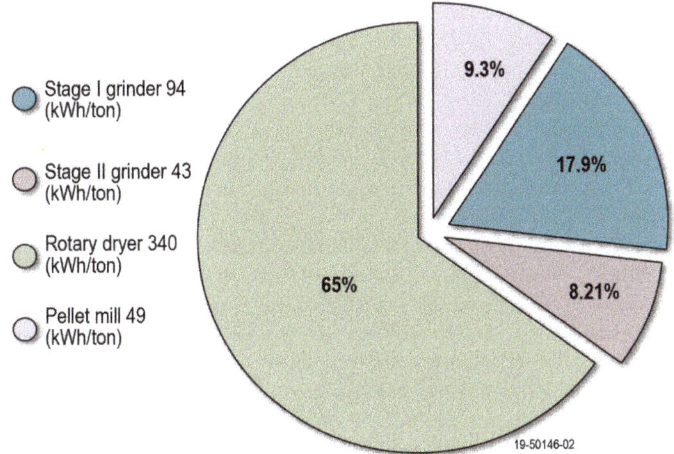

Figure 2. Energy consumption of various unit operations in pelleting of woody biomass [23].

1.3.1. Pelleting Process Variables

The pelleting process is influenced by various process variables, such as feedstock moisture, content particle size, residence time, length-to-diameter (L/D) ratio, preheating temperature, steam conditioning, and pellet die diameter [16,20]. Temperature is another important variable that can impact the quality of the produced pellets. Temperature influences the glass transition temperature of the biomass components and helps in particle bonding during pelleting. Pelleting studies conducted by Stelte et al. [31] indicated pelleting pressure decreased significantly at 100 °C for woody biomass. Compression pressure, which is dependent on the L/D ratio of the pellet die, also influences the quality of the produced pellets. Under steady state pellet production, the compression pressure equals to the extrusion pressure. If the compression is lower than the extrusion pressure, it generates back pressure and blocks the die. According to Holm et al. [32], pelleting pressure is dependent on the friction coefficient, the compression ratio, and the Poisson ratio. Their model does not take moisture content, particle size, and preheating temperature into account, but the Poisson ratio and friction coefficient for given biomass is dependent on the biomass type, temperature, and moisture content.

Feedstock composition is another important variable that impacts pellet quality and the consumption of energy during the process. In general, the degradation of hemicellulose produces adhesive products that result in natural bonding. Additionally, lignin in the biomass helps to form the solid bridges above the glass transition temperature and helps to form densified products. Van Dam et al. [33] noted that lignin above 140 °C acts as a binder and helps to form more durable pellets. Tumuluru et al. [16], Pradhan et al. [20], Serrano et al. [34], Mani et al. [35], Shaw et al. [36], Carone et al. [37], and Puig-Arnavat et al. [38] have all indicated that moisture plays a major role in the quality of the produced pellets. Most of the pelleting work reported in the literature is in the 10–15% (w.b.) moisture content. Higher moisture in the feedstock results in bulk density and durability loss of the produced pellets. Typically, moisture content necessary for pelleting depends on feedstock type. In general, grasses and crop residues need high moisture for pelleting as compared to woody biomass. Jackson et al. [39] noted that the pelleting of switchgrass, miscanthus, and wheat straw at 10% (w.b.) in a flat die pellet mill resulted in no pellet formation. These same authors also found that pellets with grasses formed when the moisture content was greater than 20% (w.b.). Understanding the impact of feedstock moisture content on pellet quality is very important to optimize the pelleting process in terms of pellet quality and cost.

1.3.2. High-Moisture Pelleting Process

Idaho National Laboratory (INL) has developed a high-moisture pelleting process that can help to reduce pellet-production costs significantly. Tumuluru [21–23], Tumuluru et al. [40], Bonner et al. [41], and Hoover et al. [42] have all successfully tested this process on both woody and herbaceous biomass, as well as chemically pretreated biomass, in both pilot and commercial-scale pelleting systems. In this process, the biomass is pelleted at a higher moisture content of > 15% (w.b.), while the high moisture pellets that are produced are further dried in low-temperature and low-cost dryers, such as grain or belt dryers. That is, this pelleting process replaces a rotary dryer in the front end with a grain or belt dryer at the back end of the pelleting process.

In this process, the biomass loses some moisture during pelleting due to frictional heat developed in the pellet die. Roughly 5–10% (w.b.) moisture loss in the biomass is seen during high-moisture pelleting. Also, the amount of moisture lost from the biomass depends on the initial moisture content of the feedstock. Lamers et al. [25] and Tumuluru et al. [26] indicated that by performing this process, a 40% reduction in pellet production costs could be realized as compared to the conventional method followed by industry. Even though the high moisture pelleting process is relatively new, there are some studies in the published literature where researchers have used this process to densify wood, agricultural crop residue, straws, grasses, and the compost [20,21,39,43,44].

1.4. Objectives

The overall goal of the project is to develop and demonstrate a state-of-the-art biomass merchandising and processing depot to identify and reduce sources of variation along the supply chain of two high-impact biomass species (i.e., pine and switchgrass), and to develop practices that manage biomass variability to deliver a consistent feedstock optimized for performance in a specific conversion-technology platform [5]. One way to manage the moisture and particle size variability is to densify the biomass. Many refineries are not ready to densify biomass, and the cost is a prohibitive factor. In this paper, new pelleting concepts, such as high-moisture pelleting, were tested on woody and herbaceous biomass blends, and the ability of this technology to efficiently manage the moisture in the blends was demonstrated.

Most of the pelleting work completed by earlier researchers has been focused on either woody or herbaceous biomass feedstocks. In addition, most of the reported literature on pelleting has been on the single pellet press. Harun and Afzal [45] worked on pelleting of agricultural and woody biomass blends in a single pellet press. These authors studied the effect of blending spruce and pine with reed canary grass, timothy hay, and switchgrass. Tumuluru et al. [9] and Harun and Afzal [45] indicated that pelleting agricultural biomass alone does not result in good pellet quality in terms of durability, which could be due to low lignin content in the agricultural biomass. In addition, low lignin content in the agricultural biomass results in higher energy consumption during the pelleting process.

The data on the pelleting blends of woody and herbaceous biomass at high moisture content ($\geq$ 20%, w.b.) in a continuous flat die pellet mill are not available. Experimental data on how high moisture content, the compression (L/D) ratio in the pellet die, and the blend ratio of pine and switchgrass impact pellet quality and energy consumption of the pelleting process are also not available. The present study aims to understand the pelleting characteristics of pine + switchgrass blends at high moisture contents. The specific objectives of the present study are to: (1) understand how the L/D ratio in the pellet die in a flat die pellet mill and blend moisture content in the range of 20–30% (w.b.) impact the quality of pellets produced using the blends of 2-inch top pine residue + switchgrass at different ratios (i.e., 25:75; 50:50; and 75:25); (2) develop response surface models and surface plots to understand the interactive effect of process variables on pellet quality and the specific energy consumption (SEC) of the process; and (3) optimize the response surface models to identify the process conditions that can minimize the pellet moisture content and maximize bulk density and the durability of blend pellets.

2. Material and Methods

2.1. Feedstocks

The switchgrass *P. virgatum* L. feedstock for this study was field-grown and harvested in Vonore, TN, USA, and processed with a tub grinder by Genera Energy, Inc., while 2-inch (50.8 mm) top pine residue samples were harvested from forest stands near Auburn, AL, USA. These residues were then dried, and both the 2-inch top pine residue and switchgrass were size-reduced in a hammer mill fitted with a 3/16-inch (4.76 mm) screen at Herty Advanced Biomaterials in Savannah, GA, USA [11].

2.2. Experimental Plan

In the present study, milled switchgrass + milled 2-inch top pine residue blend ratio, blend moisture content, and pellet die compression (L/D) ratio were selected as the process variables. Table 1 indicates the lower and upper limits of the process variables and the experimental plan for the pelleting tests, which were conducted using 4.76 mm (3/16-inch) grind switchgrass and 2-inch top pine residue blends. Pelleting tests were conducted at each L/D ratio in the pellet die at three different blend moisture contents. The pelleting tests were conducted at 60 Hz rotational speed of the pellet die. The diameter of the pellet die used for the present study was 6 mm. The pellets produced were used to measure physical properties, such as pellet moisture content (%, w.b.), bulk density (kg/m^3), and durability (%). LabVIEW software was used to log the power data. These data were further used to calculate the specific energy consumption of the pelleting process.

Table 1. Experimental conditions used for pelleting of milled 2-inch top pine residue and switchgrass blends.

Blend Feedstock	Process Variables	
	L/D Ratio of the Pellet Die (x_1)	**Blend Moisture Content (%, w.b.) (x_2)**
50% milled 2-inch top pine residue + 50% milled switchgrass	1.5, 2.0, 2.6	20, 25, 30
75% milled 2-inch top pine residue + 25% milled switchgrass	1.5, 2.0, 2.6	20, 25, 30
25% milled 2-inch top pine residue + 75% milled switchgrass	1.5, 2.0, 2.6	20, 25, 30

Note: Both switchgrass and 2-inch top pine residue were ground in a hammer mill fitted with a 3/16-inch (4.8 mm) screen size.

2.3. Flat Die Pellet Mill

In the present study, an ECO-10 flat die pellet mill was used to perform pelleting tests, as shown in Figure 3 [21–23,40,42,44]. Figure 4 provides a look at the various pellet dies with the different L/D ratios that were tested in this study. The pellet mill was provided with a 10 HP motor. At 60 Hz, the rotational speed of the pellet die is 350 rpm [21]. It has a hopper and screw feeder, which feed the pellet mill continuously. Flexible heating tapes are provided to the screw feeder and hopper, which can help to preheat the biomass before pelleting. A variable frequency drive is provided to the pellet mill to vary the rotational speed of the die. A horizontal pellet cooler is provided to cool the warm pellets coming out of the pellet die. Power consumption during pelleting was measured using the power meter provided to the pellet mill.

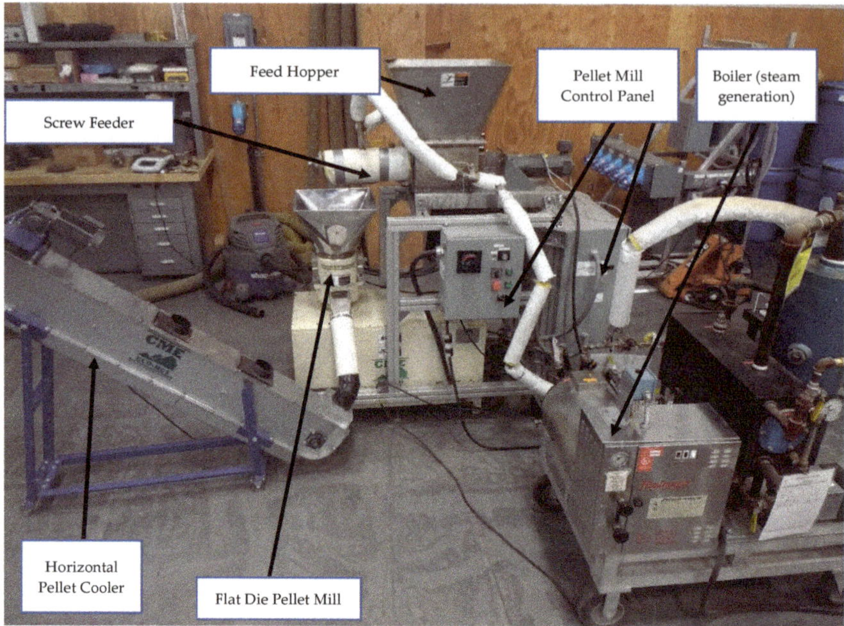

Figure 3. Flat die pellet mill at Idaho National Laboratory.

Figure 4. *Cont.*

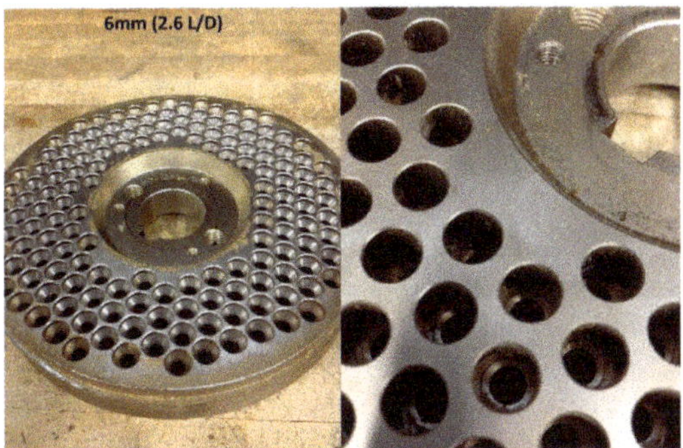

Figure 4. Flat die pellet mill dies with different length-to-diameter (L/D) ratios.

2.4. Pellet Properties Measurement

Pellet moisture content before and after drying, bulk density, and durability were measured using the American Society of Agricultural and Biological Engineers (ASABE) 2007 Standard S269.4 [46]. A complete description of these methods was given by Tumuluru [21–23] and Tumuluru et al. [40]. In the case of moisture content, the biomass is dried in a mechanical oven at 105 °C for 24 hours. In the case of bulk density, the dried pellets were poured in a cylindrical container, and the excess material was removed by striking a straight edge across the top of the container. The weight of the pellets in the container divided by the volume of the container gives the bulk density. Pellet durability was measured using the pellet durability tester, which has four compartments. Pellet samples were placed in each compartment and then rotated at 50 rpm for 10 min. The ratio of the mass of the pellets after tumbling to the mass of the pellets before tumbling is defined as pellet durability. All pellet properties are measured in triplicates. Power-consumption data during pelleting were logged using LabVIEW software (2010 Professional Service Pack 1, National Instruments, 11500 N Mopac Expwy, Austin, TX, USA) [21]. An APT power-monitor meter (NK Technologies, San Jose, CA, USA) connected to the pellet mill records the power in kilowatts. The no-load power at 60 Hz rotational speed was recorded by running the pellet mill empty. Specific energy consumption was calculated by subtracting the no-load kW from the full-load power using Equation (1).

$$Specific\ energy\ consumption = \frac{(Full\ load\ power\ (kW) - No\ load\ power\ (kW)) * time\ (h)}{weight\ of\ biomass\ processed\ (ton)} = \frac{kW \cdot h}{ton} \tag{1}$$

2.5. Response Surface Analysis

Experimental data was used to develop the second-order response model Equation (2).

$$y = b_0 + \sum_{i=1}^{n} b_i x_i + \sum_{i=1}^{n} b_{ii} x_i^2 + \sum_{i=1}^{n} \sum_{j=1}^{n} b_{ij} x_i x_j + \varepsilon \tag{2}$$

where:

- x_i and x_j are independent variables
- y is the dependent variable
- b_0, b_i, and b_j are constants
- n is the number of independent variables
- ε is an unobservable error.

Surface plots were further developed using the models. These surface plots were drawn to understand the interactive effect of the process variables on pellet quality and specific energy consumption. Statistica software, version 9.1 (StatSoft. Inc., 2300 East 14th St. Tulsa, OK, USA (www.statsoft.com)), was used to do the response surface analysis [47].

In the case of optimization, the response surface models (Equations (3)–(5)) that were developed using the experimental data were then further used as the objective functions. These objective functions are either minimized or maximized using the hybrid genetic algorithm developed by Tumuluru and McCulloch [48]. As the genetic algorithm is heuristic in nature, and they seldom reach global optimum. So, for better conversion of the optimum values, Tumuluru and McCulloch [48] hybridized a genetic algorithm with a gradient-based method. These authors tested the algorithm on food and bio-engineering problems and concluded that the hybrid genetic algorithm has better convergence compared to a regular genetic algorithm. In the case of bulk density and durability, the objective functions were maximized, whereas, in the case of pellet moisture content, the objective function is minimized to find the optimum process conditions.

$$f(y) = Minimize \ (PMC \ model) \tag{3}$$

$$f(y) = Maximize \ (BD \ model) \tag{4}$$

$$f(y) = Maximize \ (D \ model) \tag{5}$$

Note: PMC: pellet moisture content (%, w.b.); BD: bulk density (kg/m^3); D: durability (%).

3. Results

3.1. Physical Properties of the Milled 2-Inch Top Pine Residue and Milled Switchgrass Blends

Edmunds et al. [11] discussed the physical properties of 2-inch and 6-inch top pine residue, switchgrass, and blends of 2 and 6-inch top pine residue and switchgrass in detail. The average values of milled switchgrass, milled 2-inch top pine residue and blends particle size information, bulk, particle and tap densities, compressibility, and Hausner ratio are given in Table 2. The bulk and tapped density of the blends of the 2-inch top pine residue + switchgrass indicated that d50 increased with an increase in pine percentage, whereas the span reduced with an increase in the pine percentage. The trends were similar for bulk, tap, and particle densities where higher pine percentage increased the density values. Edmunds et al. [11] reasoned that a higher span of switchgrass particles could be due to an elongated nature and a higher aspect ratio of the switchgrass grind. Also, the elongated nature of the switchgrass particles can result in entanglements of the particle, which increase void spaces and reduce density. The calculated flow properties, such as the Hausner ratio, were calculated using physical properties data. The Hausner ratio of the blends was in a range between 1.26 and 1.33 [11].

Table 2. Physical characteristics of grinds from switchgrass and 2-inch top pine residue blends [11].

	Milled Switchgrass	Milled 2-inch Top Pine Residue	d50 (μm)	Span	Bulk Density (kg/m^3)	Particle Density (kg/m^3)	Tap Density (kg/m^3)	CM (%)	HR
	1	0	534	2.12	166.0	1443.9	210.3	10.6	1.26
	0	1	811	1.64	231.1	1439.7	301.7	11.9	1.31
Mass Fraction	0.75	0.25	674	2.07	180.4	1428.3	227.7	10.8	1.26
	0.50	0.50	766	2.01	188.1	1427.6	240.3	10.6	1.28
	0.25	0.75	801	1.97	207.6	1417.9	270.0	10.6	1.30

Note: CM: compressibility; and HR: Hausner ratio.

Figure 5 shows the pellets made from 2-inch top pine residue + switchgrass blends at different blend moisture contents, and L/D ratio in the pellet die. The pelleting experiments were conducted based on the experimental design provided in Table 1. Some of the key results of blending 25% switchgrass + 75% 2-inch top pine residue, 75% switchgrass+25% 2-inch top pine residue, and 50%

switchgrass + 50% 2-inch top pine residue that helped to achieve the durability and bulk density (> 95% and > 550 kg/m^3) are provided in Figures 6 and 7. The pelleting process conditions that resulted in bulk density > 550 kg/m^3 and durability > 95% were an higher L/D ratio of 2.6 and a blend moisture content of 20% (w.b.) for all the three blend ratios tested.

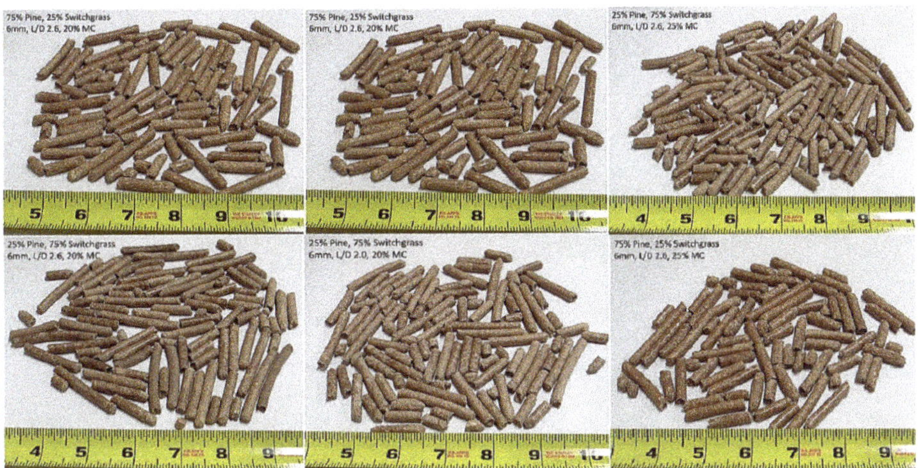

Figure 5. Blend pellets made using milled 2-inch top pine + milled switchgrass blends at different moisture contents and L/D ratios of the pellet die.

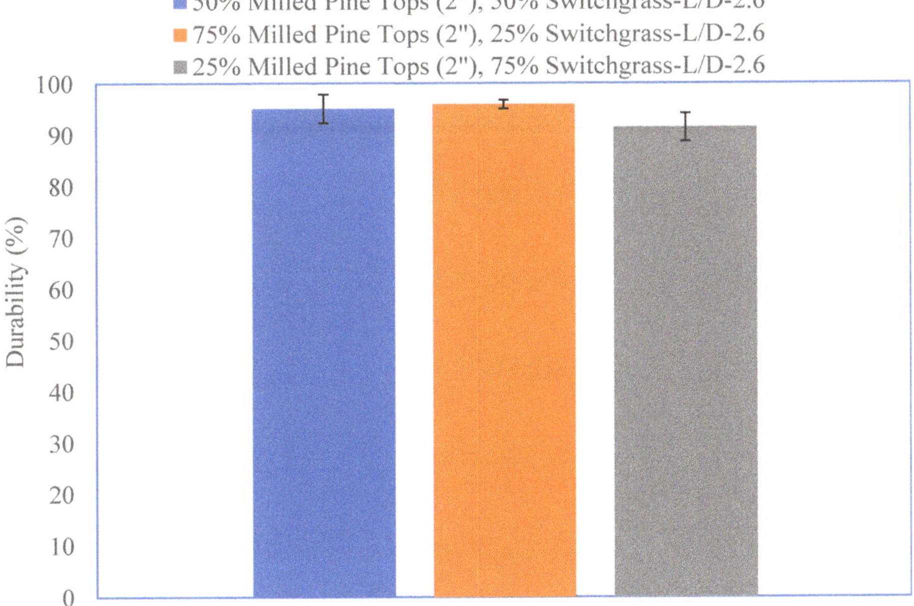

Figure 6. Durability of the pellets produced using a blend of milled 2-inch top pine residue + milled switchgrass.

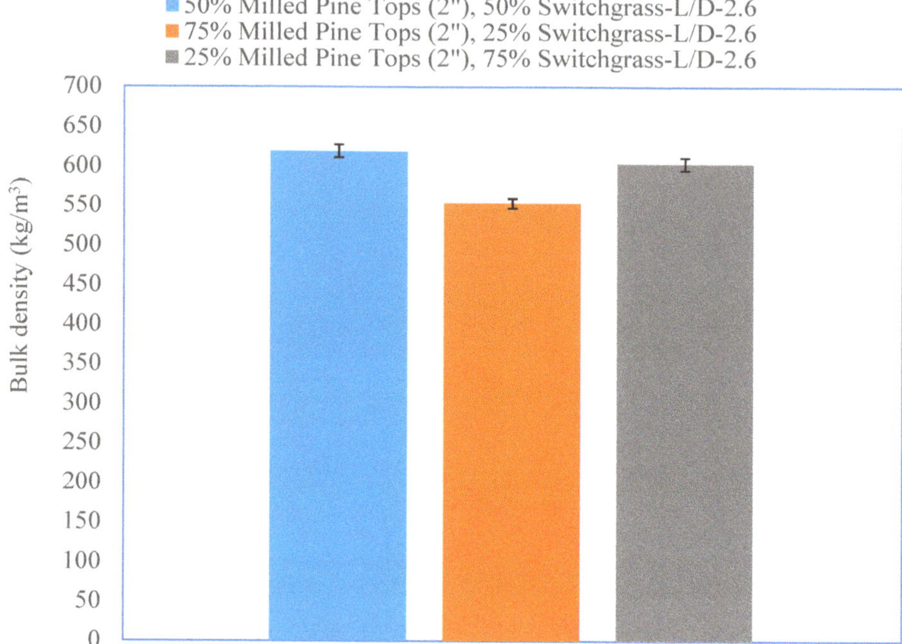

Figure 7. Bulk density of the pellets produced using a blend of milled 2-inch top pine residue + milled switchgrass.

3.2. Response Surface Models and Plots

Table 3 indicates the models developed for the blends of 2-inch top pine residue + switchgrass based on the experimental data obtained. Coefficient-to-determination values, which were in the range of 0.60 to 0.98, suggest that the models have described the pelleting process reasonably well with respect to the process variables tested. The statistical significance of the models developed for these different blends was evaluated based on their p values. For the 50% 2-inch top pine residue +50% switchgrass, the models developed pellet moisture content, bulk density and specific energy consumption were found to be statistically non-significant, whereas for durability it was found to be significant ($p < 0.01$). In the case of the 75% 2-inch top pine residue + 25% switchgrass, the models developed for pellet moisture content and durability were as found to be statistically non-significant, whereas the bulk density and specific energy consumption were found to be statistically significant at $p < 0.05$. Finally, in the case of the 25% 2-inch top pine residue + 75% switchgrass blend models, the pellet moisture content, bulk density, and durability were found to be statistically significant ($p < 0.05$, $p < 0.01$), whereas the specific energy consumption model was found to be statistically non-significant. Using these equations, response surface plots were developed. The significance of response surface plots is that they assist in understanding the interactive effect of the process variables (i.e., 2-inch top pine residue + switchgrass blend, blend moisture content, and compression ratio or L/D ratio of the pellet die) on product quality (i.e., blend pellet moisture content, bulk density, and durability) and the specific energy consumption of the pelleting process.

Table 3. Response surface models describing pellet properties and energy consumption of blends in respect to process conditions.

Physical Properties and Specific Energy Consumption	Equation	(R^2)
Blend Ratio: 50% milled 2-inch top pine residue +50% milled switchgrass		
Blend pellet moisture content (%, w.b.)	$-22.3351 + 4.5290x_1 + 2.1056x_2 + 0.1899x_1^2 -0.0190x_2^2 - 0.2090x_1x_2$	0.81
Bulk density (kg/m^3)	$511.222 - 267.332x_1 + 26.659x_2 + 98.405x_1^2 -0.614x_2^2 - 3.438x_1x_2$	0.87
Durability (%)	$4.54688 + 42.84724x_1 + 2.80468x_2 - 5.63044x_1^2 -0.04861x_2^2 - 0.38672x_1x_2$	0.97
Specific energy consumption (kWh/ton)	$5.93854 - 6.94669x_1 + 8.91335x_2 + 24.11422x_1^2 -0.09052x_2^2 - 2.87905x_1x_2$	0.73
Blend Ratio: 75% milled 2-inch top pine residue + 25% milled switchgrass		
Blend pellet moisture content (%, w.b.)	$-1.64073 + 4.10549x_1 + 0.76126x_2 - 2.69651x_1^2 -0.01661x_2^2 + 0.28656x_1x_2$	0.80
Bulk density (kg/m^3)	$248.963 - 426.649x_1 + 61.506x_2 + 93.687x_1^2 -1.641x_2^2 + 3.921x_1x_2$	0.98
Durability (%)	$127.1201 + 28.3141x_1 - 5.2011x_2 - 3.1719x_1^2 +0.0909x_2^2 - 0.2458x_1x_2$	0.60
Specific energy consumption (kWh/ton)	$157.397 - 221.180x_1 + 11.486x_2 + 44.171x_1^2 -0.233x_2^2 + 1.273x_1x_2$	0.88
Blend Ratio: 25% milled 2-inch top pine residue + 75% milled switchgrass		
Blend pellet moisture content (%, w.b.)	$71.3611 - 24.2603x_1 - 3.2983x_2 + 5.4860x_1^2 +0.0749x_2^2 + 0.1095x_1x_2$	0.98
Bulk density (kg/m^3)	$-390.985 + 402.475x_1 + 49.246x_2 - 71.907x_1^2 -1.162x_2^2 - 1.831x_1x_2$	0.97
Durability (%)	$96.55725 - 0.74423x_1 - 1.16415x_2 - 1.00586x_1^2 -0.01055x_2^2 + 0.60131x_1x_2$	0.88
Specific energy consumption (kWh/ton)	$-650.876 + 371.666x_1 + 37.340x_2 - 87.743x_1^2 -0.874x_2^2 + 0.821x_1x_2$	0.66

Note: Both switchgrass and 2-inch top pine residue were ground in a hammer mill fitted with a 3/16-inch (4.8 mm) screen size; x_1: L/D ratio of the pellet die; x_2: Blend moisture content (%, w.b.).

3.2.1. Blend Ratio: 50% 2-inch Milled Pine Top Residue + 50% Milled Switchgrass

The moisture content of the pellet decreased with a decrease in blend moisture content. The lowest pellet moisture content of < 14% (w.b.) was observed at an L/D ratio of 1.5 to 2.6 for 20% (w.b.) blend moisture content tested, as shown in Figure 8. The L/D ratio of the die did not have a significant effect on moisture loss during pelleting. This observation corroborated with earlier studies on a pilot-scale pellet mill for corn stover feedstock [21–23]. The bulk density of the blend pellets decreased at higher blend moisture content and lower L/D ratio in the pellet die. Higher bulk densities of > 580 kg/m^3 were observed at an L/D ratio of 2.6 and a blend moisture content of about 20–22% (w.b.), as shown in Figure 9. In the case of pellet durability, the L/D ratio had a significantly greater effect than blend moisture content. A lower to medium moisture content of 20–25% and a higher L/D ratio of 2.6 resulted in durability values of > 94%, as shown in Figure 10. Specific energy consumption decreased with a decrease in the L/D ratio in the pellet die. A higher L/D ratio in the pellet die (about 2.4–2.6) and a lower moisture content of 20–22% (w.b.) resulted in a higher specific energy consumption of > 140 kWh/ton, whereas lowering the L/D ratio to 1.5 at the same moisture content resulted in a lower specific energy consumption of < 102 kWh/ton, as shown in Figure 11.

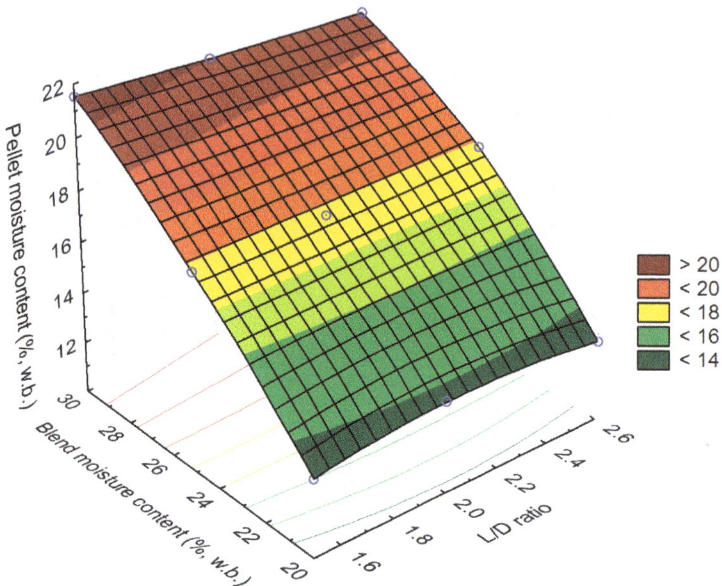

Figure 8. Effect of blend moisture and L/D ratio on pellet moisture content of 50% 2-inch top pine residue + 50% switchgrass.

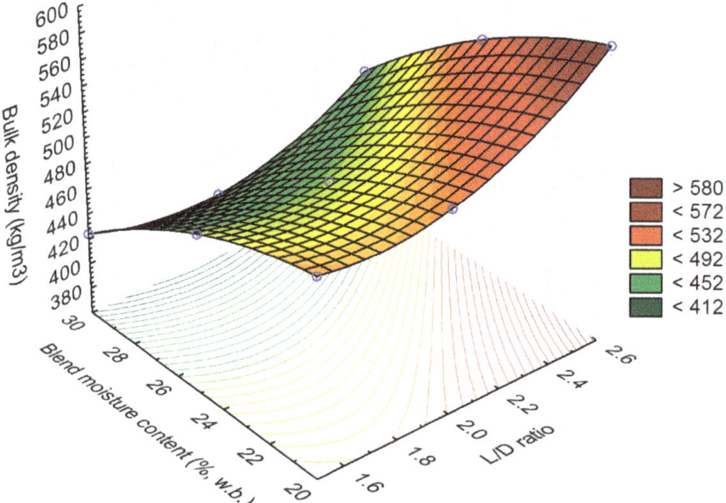

Figure 9. Effect of blend moisture and L/D ratio on pellet bulk density of 50% 2-inch top pine residue + 50% switchgrass.

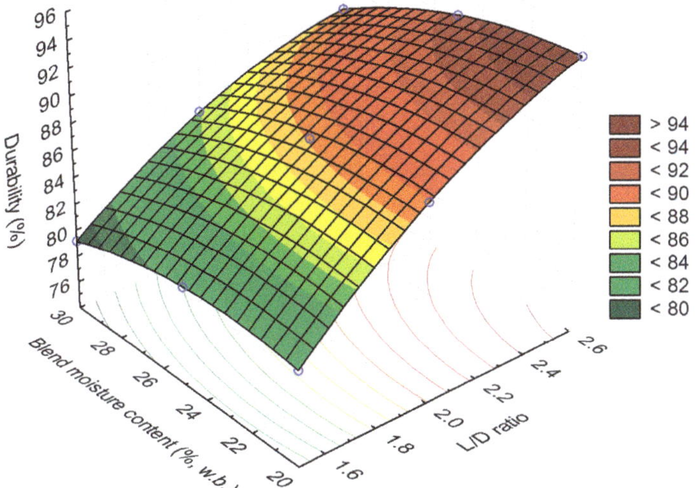

Figure 10. Effect of blend moisture and L/D ratio on pellet durability of 50% 2-inch top pine residue + 50% switchgrass.

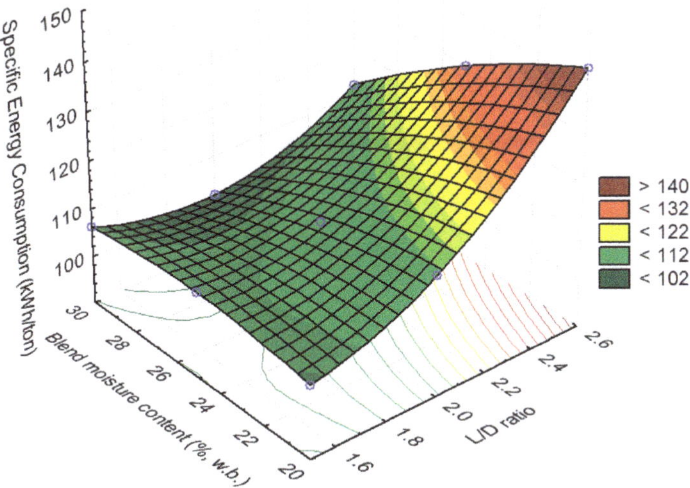

Figure 11. Effect of blend moisture and L/D ratio on specific energy consumption of 50% 2-inch top pine residue + 50% switchgrass.

3.2.2. Blend Ratio: 75% 2-inch Milled Pine Top Residue and 25% Milled Switchgrass

Lowering the blend moisture content and increasing the L/D ratio decreased pellet moisture content. An L/D ratio of 2.6 and a lower blend moisture content of 20% (w.b.) resulted in pellets with a moisture content of < 14.5% (w.b.), whereas the same moisture content with a lower L/D ratio resulted in higher moisture content in the blended pellets (16.5%), as shown in Figure 12. Bulk density of the blended pellet increased with an increase in the L/D ratio in the pellet die (to 2.6) and a decrease in blend moisture content to 20% (w.b.). The highest and lowest bulk density values observed were > 540 and < 364 kg/m^3, as shown in Figure 13. The durability values of the produced pellets using pine and switchgrass blends were positively influenced by the L/D ratio but negatively influenced by the moisture content of the blends. Increasing the L/D ratio to 2.6 and decreasing the moisture content to 20% produced pellets with durability values of > 98%, while increasing the moisture content to 30% and decreasing the pellet die L/D ratio to 1.5 reduced the durability values to < 78%, as observed in Figure 14. The lower specific energy consumption of < 78 kWh/ton was observed for the L/D ratio of 2.2 to 2.6 at a lower blend moisture content of 20% (w.b.), as shown in Figure 15.

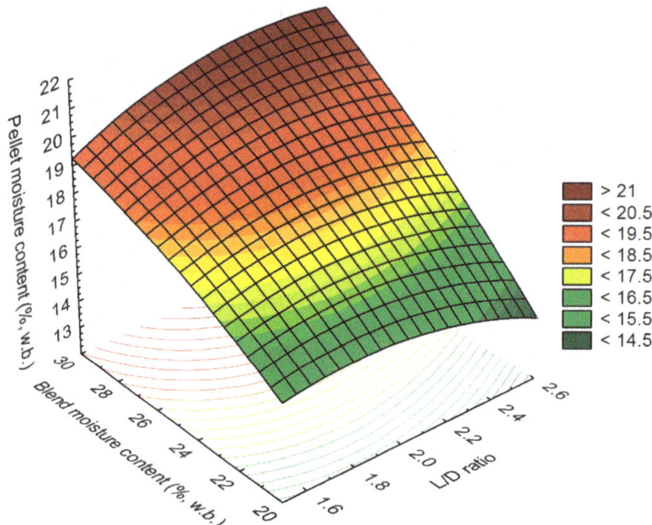

Figure 12. Effect of blend moisture and L/D ratio on pellet moisture content of 75% 2-inch top pine residue + 25% switchgrass.

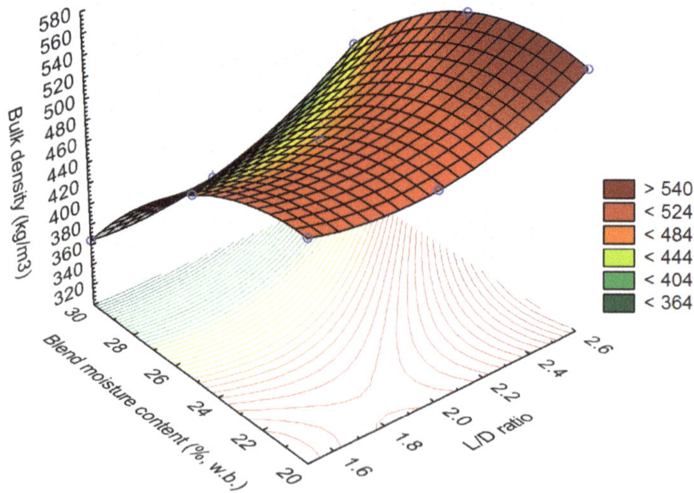

Figure 13. Effect of blend moisture and L/D ratio on pellet bulk density of 75% 2-inch top pine residue + 25% switchgrass.

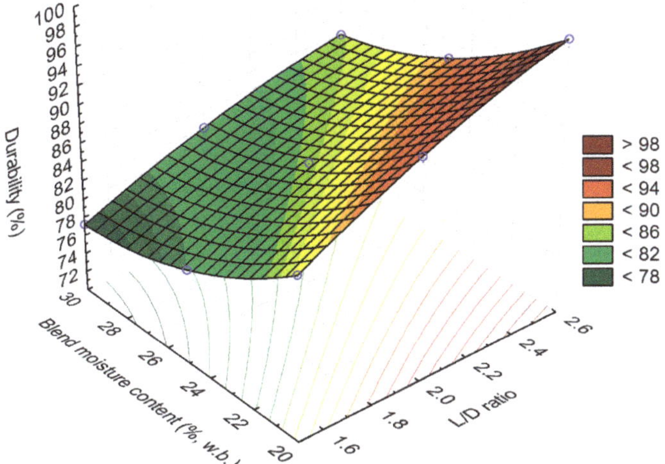

Figure 14. Effect of blend moisture and L/D ratio on durability of 75% 2-inch top pine residue + 25% switchgrass.

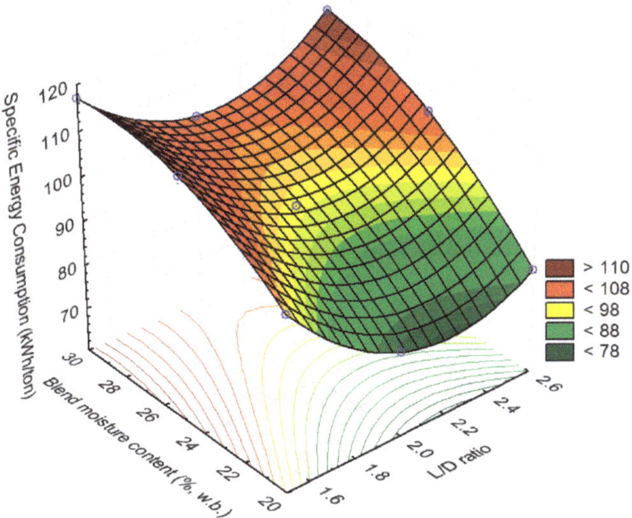

Figure 15. Effect of blend moisture and L/D ratio on specific energy consumption of 75% 2-inch top pine residue + 25% switchgrass.

3.2.3. Blend Ratio: 25% 2-inch Milled Pine Top Residue + 75% Milled Switchgrass

Lowering the blend moisture content to 20% (w.b.) and the L/D ratio to 1.6 to 2.4 reduced the pellet moisture content to < 14% (w.b.), whereas a higher moisture content in the blend at 30% (w.b.) resulted in a higher moisture content in the produced pellets, as observed in Figure 16. The loss of moisture was greater with a corresponding increase in the initial moisture content of the blend. For example, about 8–10% moisture loss was seen at 30% (w.b.) initial moisture content, whereas at 20% (w.b.) initial moisture content, the loss of moisture observed during pelleting was only about 6–7% (w.b.). The bulk density increased with an increase in the L/D ratio and a decrease in blend moisture content. The maximum bulk density observed at 20% moisture content and an L/D ratio of 2.6 was 580 kg/m^3, whereas at an L/D ratio of 1.5 and a blend moisture content of 30% (w.b.), the lowest bulk density of < 404 kg/m^3 was observed, as shown in Figure 17. The L/D ratio had a more significant effect on the durability values as compared to the blend moisture content. The maximum durability observed was > 92% at a blend moisture content of 20% and an L/D ratio of 2.6. At a lower L/D ratio of 1.5, the predicted durability values were in the range of 76–78% for the different blend moisture contents tested, as seen in Figure 18. A higher moisture content and a lower L/D ratio reduced the specific energy consumption. At an L/D ratio of 1.5 and a blend moisture content of 30% (w.b.), the specific energy consumption observed was 80 kWh/ton. At a higher L/D ratio of 2.6 and a lower moisture content of 20% (w.b.), the specific energy consumption was > 180 kWh/ton, as seen in Figure 19.

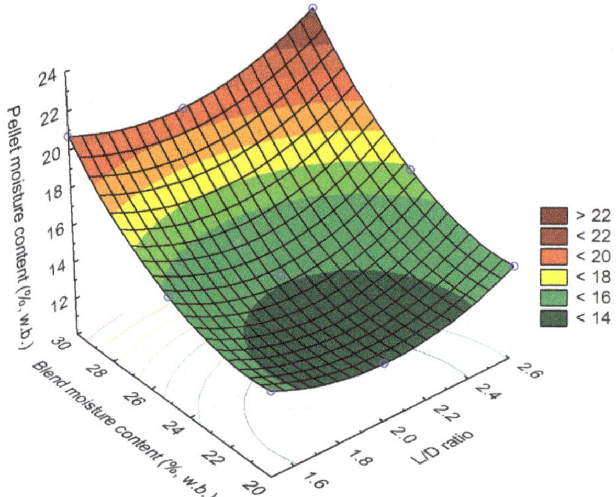

Figure 16. Effect of blend moisture and L/D ratio on pellet moisture content of 25% 2-inch top pine residue + 75% switchgrass.

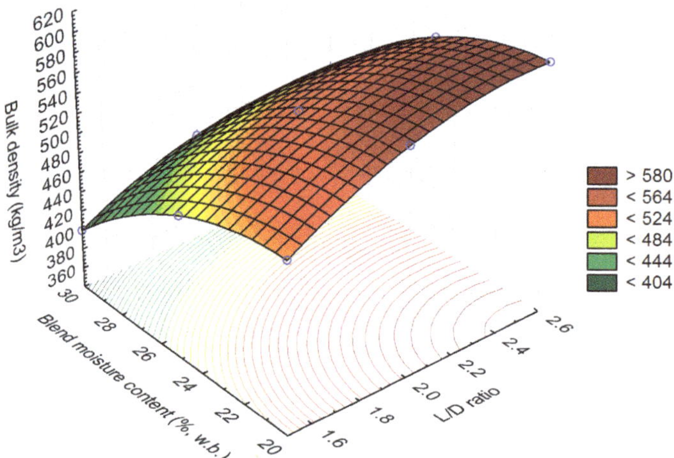

Figure 17. Effect of blend moisture and L/D ratio on pellet bulk density of 25% 2-inch top pine residue + 75% switchgrass.

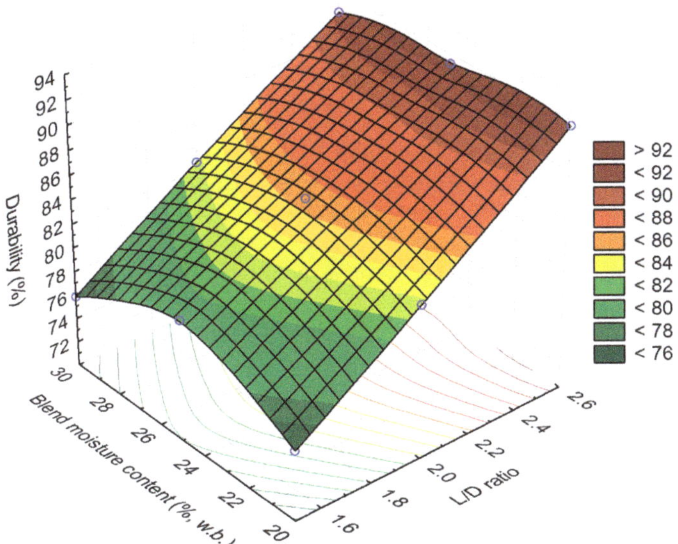

Figure 18. Effect of blend moisture and L/D ratio on durability of 25% 2-inch top pine residue + 75% switchgrass.

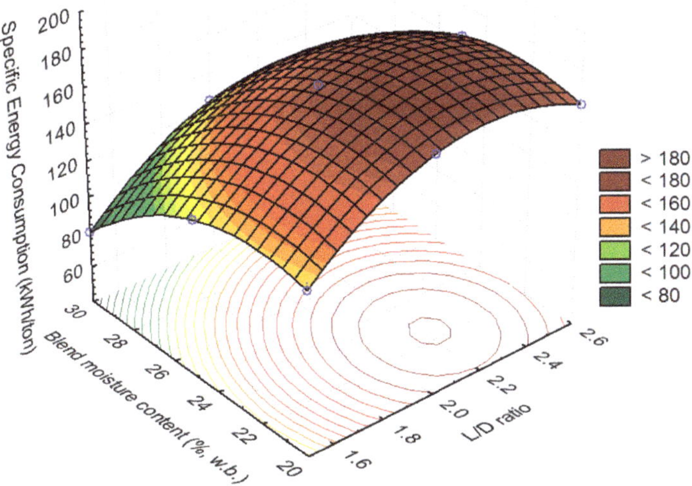

Figure 19. Effect of blend moisture and L/D ratio on specific energy consumption of 25% 2-inch top pine residue + 75% switchgrass.

3.3. Optimized Process Conditions

The regression equations given in Table 3 were further optimized to find the optimum pelleting process conditions that can maximize bulk density and durability and minimize the pellet moisture content for the different blend of ratios of pine + switchgrass tested. Tables 4–6 indicate the optimized conditions obtained using the hybrid genetic algorithm optimization developed by Tumuluru and

McCulloch [48]. It is clear from these tables that optimized process conditions identified for the three tested blends were different in terms of L/D ratio of the pellet die but lower pellet moisture content of about 20% (w.b.) was necessary to maximize bulk density and durability and minimize the pellet moisture content. For maximizing bulk density and durability, a maximum L/D ratio and minimum moisture content were desirable for all three of the blends tested. To minimize blend pellet moisture content, a lower blend moisture content was desirable for all three of the blends tested, but the L/D ratio in the pellet die was different. For 50% 2-inch top pine residue + 50% switchgrass, a lower L/D ratio of 1.58 was needed; for the 75% 2-inch top pine residue + 25% switchgrass, a higher L/D ratio of 2.55 was necessary; and for the 25% 2-inch top pine residue + 75% switchgrass, a medium L/D ratio of 2.01 was desirable to reduce the moisture content in the blended pellets. Among the three blends tested, the blend with 75% switchgrass + 25% 2-inch top pine residue and the blend with the 50% switchgrass + 50% 2-inch top pine residue produced blended pellets with durability values of about 95%, while the pellets produced with 75% 2-inch top pine residue + 25% switchgrass produced pellets with durability values of about 98.11%. In regard to bulk density, the maximum of 591 kg/m^3 was observed for the pellets produced with the 25% 2-inch top pine residue + 75% switchgrass, whereas the other two combinations produced pellets with durability values of 583 kg/m^3 and 554 kg/m^3.

Table 4. Optimum process conditions for maximizing density and durability and minimizing pellet moisture content for 50% milled 2-inch top pine residue + 50% milled switchgrass blends.

Pellet Properties	Maximum	Minimum	Individual Optimum Process Conditions	
	-	-	L/D Ratio of Pellet Die	Blend Moisture Content (%, w.b.)
Pellet Moisture Content (%, w.b.)	-	13.25	1.58	20.002
Bulk Density (kg/m^3)	583.57		2.56	20.002
Durability (%)	93.94		2.52	20.23

Table 5. Optimum process conditions for maximizing bulk density and durability and minimizing pellet moisture content for 75% milled 2-inch top pine residue + 25% milled switchgrass blends.

Pellet Properties	Maximum	Minimum	Individual Optimum Process Conditions	
	-	-	L/D Ratio of Pellet Die	Blend Moisture Content (%, w.b.)
Pellet Moisture Content (%, w.b.)	-	14.625	2.55	20.19
Bulk Density (kg/m^3)	554.00	-	2.58	21.78
Durability (%)	98.11	-	2.54	20.16

Table 6. Optimum process conditions for maximizing bulk density and durability and minimizing pellet moisture content for 25% milled 2-inch top pine residue + 75% milled switchgrass blends.

Pellet Properties	Maximum	Minimum	Individual Optimum Process Conditions	
	-	-	L/D Ratio of Pellet Die	Blend Moisture Content (%, w.b.)
Pellet Moisture Content (%, w.b.)	-	13.13	2.01	20.66
Bulk Density (kg/m^3)	591.55	-	2.34	20.02
Durability (%)	91.53	-	2.59	20.20

4. Discussion

Moisture loss was observed when 2-inch top pine + switchgrass blends that were pelleted at high mositure contents. The loss of moisture varied for the blend ratios tested and for the pelleting process variables, such as the L/D ratio and blend moisture content. There was about 6–10% (w.b.) moisture loss during pelleting, and the loss was largely dependent on the initial moisture content of the blend, and less on the L/D ratio of the pellet die. This observation corroborates with earlier work [21–23] in which corn stover and lodgepole pine, during pelleting at a high moisture content, lost about 6–10% (w.b.), and the loss of moisture was dependent on the initial moisture content of the feedstock. Tumuluru et al. [23] has reasoned that during pelleting, the mositure loss in the biomass is due to: (a) mositure flash-off due to the frictional heat developed in the die; and (b) cooling. This leads to drying most of the pellet surface moisture, resulting in partially dried pellets. Also, it is important to dry the partially dried pellets slowly; otherwise, it can result in case-hardening, making pellets harder outside but trapping moisture inside, which can also result in microbial degradation during storage. Tumuluru [44] indicated that the high moisture pelleting process not only densifies the biomass, but helps to drive some of the moisture from the feedstock. Also, high-moisture pelleting makes drying optional. If, for example, the pellets do not have to be stored for long periods of time and do not require transportation over longer distances, the partially dried pellets can be used as such without any further drying for the biochemical conversion process. This is generally true in biochemical conversion where biomass is rewetted during pretreatment and conversion. Also, in the high moisture pelleting process, the moisture in the biomass is more efficiently managed, which reduces the cost of preprocessing significantly. Lamers et al. [25] indicated a 40% reduction in pellet production costs mainly due to moisture loss during pelleting and drying the high moisture pellets using low-temperature dryers, such as grain or belt dryers, provide cost-savings that are 10 times lower and can operate using low quality heat.

In general, low bulk density is another major limitation of herbaceous biomass and results in issues related to storage, handling, and transportation [16,49,50]. These limitations pose a serious challenge for biomass applications on a commercial scale. The present pelleting study indicates that bulk density increases by almost 3–5 times over the raw material, and the increase in the density is dependent on the process conditions selected. In their studies on biomass blending and densification impacts on the feedstock supply chain and biochemical conversion, Ray et al. [10] concluded that low-density biomass requires more resources for transportation and shipping. In their review on biomass densification systems, Tumuluru et al. [16] suggested that pellet mills, briquette presses, cubers, agglomerators, and tablet presses all help to improve bulk density and produce a consistent product in terms of physical properties (e.g., size, shape, bulk density). Densification of biomass also helps to improve handling and conveyance efficiencies in biomass supply systems and infeed [16]. A big challenge for using biomass blends in biorefineries is feeding and handling. Due to variations in bulk density and particle size distribution, the blends will segregate during storage, handling, and feeding, and can influence feed-handling and conversion-process efficiencies. According to Ray et al. [10], the use of blended and densified feedstocks in conversion pathways instead of conventionally ground biomass from a single source addresses several challenges in the current biomass supply chain, such as transportation, storage, cost, quality, and supply variability. Edmunds et al. [11], Sahoo and Mani [50], and Tumuluru et al. [51] reported that herbaceous biomass, such as switchgrass, has a bulk density in the range of 150–160 kg/m^3. Based on the present study, pelleting blends of switchgrass + 2-inch top pine residue increased bulk density values to about 540–580 kg/m^3. Also, because the moisture content of the pellets is < 10% (w.b.), they are more aerobically stable during storage.

The present research indicated that both the L/D ratio of the pellet die or compression pressure and blend moisture content influenced the bulk density and durability of the produced pellets. A higher L/D ratio and lower moisture content increased the bulk density for all the blend ratios tested. Studies conducted by Said et al. [52] on rice straw in a flat die pellet mill showed that the durability of the pellets is strongly dependent on the effectiveness of the interparticle bonds created during pelleting. Their studies indicated that higher moisture content (10–17%, w.b.) increased durability, but decreased bulk density values. A similar observation was observed by Serrano et al. [34] on barley straw, where an increase in moisture content increased the length of the pellet and its durability but decreased durability values. Studies conducted by Rhén et al. [53] on the pelleting of woody biomass (*Norway spruce*) at different preheating temperatures and pressure indicated that both preheating temperature and moisture content had a significant effect on the bulk density of the pellets produced. Studies conducted by Jackson et al. [39] and Sarkar et al. [54] also indicated that pelleting corn stover and switchgrass at a higher moisture content of about 20–26% (w.b.) resulted in pellets with a bulk density in the range of 500–600 kg/m^3. The research conducted by the earlier researcher and the observations from the present study also seems to corroborate that increasing the moisture content decreases the bulk density of the pellets produced.

Currently, the major challenge to use pelleted biomass in biorefining operations is the cost. In this study, the high moisture pelleting process that was tested helps to significantly reduce pelleting costs. Also, this process helps to produce pellets with different bulk density and durability values. According to Tumuluru [21], if pellets are transported by a truck, which is a weight-limited system, very high bulk densities are not needed to fill the truck. Based on maximum weight and volume of the truck, densified products with a bulk density in the range of 350–400 kg/m^3 can fill the truck to capacity. Also if the pellets are transported to shorter distances they do not neet to meet the durability standards set for long-distance transportation. Tumuluru [21] suggested that the cost of pellet production using conventional method cannot be completely offset by saving in the transportation costs especially if the transportation distances are less than 200–300 miles. One way to make pelleting an economically viable technology for the biorefineries is by reducing the cost. The high-moisture pelleting tested in this study can make pelleting more cost-effective. Also, the cost savings achieved in terms of storage, handling, and feeding due to the use of pellets are not quantified throughly, it they are quantified it can make pelleting a more favorable operation for biorefineries. Another major advantage of blending woody with herbaceous biomass is that it improves the chemical composition. Woody biomass has a higher carbon content and is lower in ash, while the herbaceous biomass is lower in carbon content and higher in ash. Blending woody with herbaceous biomass can help to overcome herbaceous biomass feedstock specification limitations and make them meet specifications required for thermochemical conversion in terms of calorific value, volatiles, oxygen, hydrogen, nitrogen, chlorine, sulfur, nitrogen, and ash content [9].

In general, the lignin in the biomass is considered a natural binding agent and plays an important role in the densification process. In the present study, increasing the pine content in the blend to 75% increased durability values and reduced the specific energy consumption. In his studies on the pelleting of woody and herbaceous biomass at high moisture content, Tumuluru [44] indicated that higher lignin content in woody biomass increased the bulk and durability values of the pellets. In addition, the same study also indicated that energy sorghum resulted in low-quality pellets in terms of their density and durability. According to Tumuluru et al. [9], grasses with lower lignin content are difficult to pelletize and consume higher pelleting energy, in addition to producing low-quality pellets in terms of their density and durability. However, the same authors indicated that blending straws and grasses with woody biomass, which has higher lignin and lower ash content, could help to improve pellet properties and reduce pelleting energy consumption. In their studies on the chemical and mechanical propeties of agricultural and woody biomass, Harun and Afzal [45] indicated that higher percentages of woody biomass in the blend of pine and switchgrass increased the pellet strength and durability values. This present research corroborates this observation and proves that blending pine with switchgrass does indeed help to produce a good quality of pellet in terms of durability and reduce specific energy consumption.

Edmunds et al. [11] indicated that switchgrass has about 21% lignin content, while 2-inch top pine residue has about 37.5% on an as-received dry-weight basis. The previous research published on pelleting of grasses indicated that grasses take more energy to pellet as well as they do not make a good pellet due to its low lignin content and needle-shaped particles. Pine and switchgrass blending studies conducted by Edmunds et al. [11] indicated that significant improvement in terms of lignin content and particle size distribution could be achieved. These improvements in terms of physical properties and biochemical composition, especially lignin, can help to produce good quality pellets at lower energy consumption. The blending of these types of biomass not only helps to modify their chemical composition but often improve their pelleting characteristics as well, due to better interlocking ability and flowability of the biomass in the pellet die. This observation was corroborated by the present study, where increasing the pine percentage to 75% in the blend improved the durability of the pellets. Also, the energy consumption of the pelleting process was lower when the pine percentage was higher in the blends tested. The improvements in bulk density and durability and lower energy consumption for the pine and switchgrass blend pellets tested can be due to improved chemical composition and particle size distribution, which might have resulted in better flow characteristics in the pellet die.

Many researchers have indicated that particle size distribution has a significant impact on the quality of the produced pellets [16,55,56]. It is critical to manage the particle size to produce the right quality of densified products at a lower specific energy consumption. The blending of woody and herbaceous biomass helps to alter particle size distribution and can make feedstock suitable for different densification systems. In general, a pellet mill requires smaller particles as the contact area between the particles plays a major role in creating necessary bonding between the particles. The common bonding mechanism during pelleting are: (1) particle bonding due to interfacial forces and capillary pressures; and (2) solid bridges that are formed due to chemical reactions, sintering, solidification, hardening of the binder, hardening of the melted substances, or crystallization of the dissolved materials results in agglomeration of biomass particles [16]. In addition, according to MacBain [55] and Payne [56], finely ground materials are suitable for pelleting because they have higher surface area to absorb steam during conditioning and can result in higher-starch gelatinization and increased particle binding. The same authors have also suggested that a certain ratio of fines—medium and coarse particles—are necessary to improve pellet quality and reduce pelleting energy consumption. Based on the present study, blending of pine and switchgrass at different ratios might have influenced particle size distributions, positively impacted pellet quality (i.e., bulk density and durability), and reduced the overall specific energy consumption of the pelleting process. Future work on the pelleting of woody and herbaceous biomass blends should be focused on testing the process in a ring die pellet mill at both the pilot and commercial scales; understanding how the chemical composition and energy properties changes with respect to moisture content and pelleting process variables, such as the L/D ratio; and understanding the effect of grind size on the quality of the pellets and energy consumption of the process.

5. Conclusions

Based on the findings in the present research, the following conclusions were drawn:

1. Moisture loss during pelleting was higher at high blend moisture content, which corroborates with earlier studies on pelleting corn stover using a laboratory-scale felt die and a pilot-scale ring die pellet mill. There is about 6–10% (w.b.) moisture loss during pelleting of blends of 2-inch top pines residues and switchgrass blends.
2. Both blend moisture content and L/D ratio of the pellet die had a significant effect on the pellet properties and the specific energy consumption for the three blend ratios tested.
3. All three blend ratios produced pellets with bulk density values > 550 kg/m^3. Increasing the 2-inch top pine residue percentage in the blend improved pellet physical properties.
4. The 75% 2-inch top pine residue + 25% switchgrass combinations helped to produce pellets with durability values of > 95%, whereas 50% 2-inch top pine residue + 50% switchgrass produced pellets with durability values closer to 95%.
5. Higher switchgrass percentage increased the specific energy consumption of the pelleting process. At a higher 2-inch pine residue percentage, the specific energy consumption reduced to about 90 kWh/ton, whereas increasing the moisture content of the blend ratio increased the specific energy consumption.

6. The future work should be focused to understand the pelleting characteristics of high mositure pine residue, switchgrass, and blends of pine and switchgrass in a pilot scale ring die pellet mill. Evaluate how the pelleting process variables impact the physical properties, proximate and ultimate composition, and energy properties.

Acknowledgments: This work was supported by the U.S. Department of Energy, Office of Energy Efficiency and Renewable Energy under DOE Idaho Operations Office Contract DE-AC07-05ID14517. Accordingly, the U.S. Government retains and the publisher, by accepting the article for publication, acknowledges that the U.S. Government retains a nonexclusive, paid-up, irrevocable, worldwide license to publish or reproduce the published form of this manuscript or allow others to do so, for U.S. Government purposes. The author would like to thank the project partners Genera Bioenergy, Vonore, TN, University of Tennessee, Knoxville, TN Auburn University, Auburn, AL and Herty Advanced Biomaterials, Savannah, GA, USA for providing 3/16-inch grind switchgrass and 2-inch top pine residue samples for pelleting tests. The author also wants to thank Matt Dee, Research Engineer and Craig. C. Conner, retired INL employee, for supporting the experimental work.

Conflicts of Interest: The authors declare no conflict of interest.

U.S. Department of Energy Disclaimer: This information was prepared as an account of work sponsored by an agency of the U.S. Government. Neither the U.S. Government nor any agency thereof, nor any of their employees, makes any warranty, express or implied, or assumes any legal liability or responsibility for the accuracy, completeness, or usefulness of any information, apparatus, product, or process disclosed, or represents that its use would not infringe privately owned rights. References herein to any specific commercial product, process, or service by trade name, trademark, manufacturer, or otherwise, does not necessarily constitute or imply its endorsement, recommendation, or favoring by the U.S. Government or any agency thereof. The views and opinions of authors expressed herein do not necessarily state or reflect those of the U.S. Government or any agency thereof.

References

1. U.S. Department of Energy, Office of Energy Efficiency and Renewable Energy. Bioenergy Basics. 2018. Available online: https://www.energy.gov/eere/bioenergy/bioenergy-basics (accessed on 30 January 2019).
2. Langholtz, M.H.; Stokes, B.J.; Eaton, L.M. *Billion-Ton Report: Advancing Domestic Resources for a Thriving Bioeconomy, Volume 1: Economic Availability of Feedstocks*; Oak Ridge National Laboratory: Oak Ridge, TN, USA, 2016.
3. Perlack, R.D.; Wright, L.L.; Turhollow, A.; Graham, R.L.; Stokes, B.J.; Erbach, D.C. *Biomass as Feedstock for a Bioenergy and Bioproducts Industry: The Technical Feasibility of a Billion-Ton Annual Supply*; Oak Ridge National Laboratory: Oak Ridge, TN, USA, 2005.
4. U.S. Department of Agriculture. *A USDA Regional Roadmap to Meeting the Biofuels Goals of the Renewable Fuels Standard by 2022*; USDA Biofuels Strategic Production Report; U.S. Department of Agriculture: Washington, DC, USA, 2010.
5. Rials, T.G. Next Generation Logistics Systems for Delivering Optimal Biomass Feedstocks to Biorefining Industries in the Southeastern U.S.: Logistics for Enhanced Attributes Feedstock (LEAF) DOE-Funded Project Peer Review. Available online: https://www.energy.gov/sites/prod/files/2017/05/f34/fsl_rials_123107.pdf (accessed on 30 January 2019).
6. Tumuluru, J.S. Why biomass preprocessing and pretreatments? In *Biomass Preprocessing and Pretreatments for Production of Biofuels*; Tumuluru, J.S., Ed.; CRC Press: Boca Raton, FL, USA, 2018.
7. Tumuluru, J.S. Thermal pretreatment of biomass to make it suitable for biopower application. In *Biomass Preprocessing and Pretreatments for Production of Biofuels*; Tumuluru, J.S., Ed.; CRC Press: Boca Raton, FL, USA, 2018.
8. Yancey, N.A.; Tumuluru, J.S.; Wright, C.T. Drying, grinding, and pelletization studies on raw and formulated biomass feedstocks for bioenergy applications. *J. Biobased Mater. Bioenergy* **2013**, *7*, 549–558. [CrossRef]
9. Tumuluru, T.S.; Hess, J.R.; Boardman, R.D.; Wright, C.T.; Westover, T.L. Formulation, pretreatment, and densification options to improve biomass specifications for co-firing high percentages with coal. *Ind. Biotechnol.* **2012**, *8*, 113–132. [CrossRef]
10. Ray, A.E.; Li, C.; Thompson, V.S.; Daubaras, D.L.; Nagle, N.J.; Hartley, D.S. Biomass blending and densification: Impacts on feedstock supply and biochemical conversion performance. In *Biomass Volume Estimation and Valorization for Energy*; Tumuluru, J.S., Ed.; IntechOpen: London, UK, 2017.

11. Edmunds, C.W.; Reyes Molina, E.A.; André, N.; Hamilton, C.; Park, S.; Fasina, O.; Adhikari, S.; Kelley, S.S.; Tumuluru, J.S.; Rials, T.G.; et al. Blended feedstocks for thermochemical conversion: Biomass characterization and bio-oil production from switchgrass-pine residues blends. *Front. Energy Res.* **2018**, *6*, 79. [CrossRef]

12. Kenney, K.L.; Smith, W.A.; Gresham, G.L.; Westover, T.L. Understanding biomass feedstock variability. *Biofuels* **2013**, *4*, 111–127. [CrossRef]

13. Thompson, D.N.; Campbell, T.; Bals, B.; Runge, T.; Teymouri, F.; Ovard, L.P. Chemical preconversion: Application of low-severity pretreatment chemistries for commoditization of lignocellulosic feedstock. *Biofuels* **2013**, *4*, 323–340. [CrossRef]

14. Crawford, N.C.; Ray, A.E.; Yancey, N.A.; Nagle, N. Evaluating the pelletization of "pure" and blended lignocellulosic biomass feedstocks. *Fuel Process. Technol.* **2015**, *140*, 46–56. [CrossRef]

15. Mahadevan, R.; Adhikari, S.; Shakya, R.; Wang, K.; Dayton, D.; Lehrich, M.; Taylor, S.E. Effect of alkali and alkaline earth metals on in-situ catalytic fast pyrolysis of lignocellulosic biomass: A microreactor study. *Energy Fuels* **2016**, *30*, 3045–3056. [CrossRef]

16. Tumuluru, J.S.; Wright, C.T.; Hess, J.R.; Kenney, K.L. A review of biomass densification systems to develop uniform feedstock commodities for bioenergy application. *Biofuel Bioprod. Biorefin.* **2011**, *5*, 683–707. [CrossRef]

17. Tumuluru, J.S.; Sokhansanj, S.; Hess, J.R.; Wright, C.T.; Boardman, R.D. A review on biomass torrefaction process and product properties for energy applications. *Ind. Biotechnol.* **2011**, *7*, 384–401. [CrossRef]

18. Tumuluru, J.S.; Yancey, N. Conventional and advanced mechanical preprocessing methods for biomass: Performance quality attributes and cost analysis. In *Biomass Preprocessing and Pretreatments for Production of Biofuels*; Tumuluru, J.S., Ed.; CRC Press: Boca Raton, FL, USA, 2018.

19. Sahoo, K.; Bilek, E.M.; Bergman, R.D.; Kizha, A.R.; Mani, S. Economic analysis of forest residues supply chain options to produce enhanced-quality feedstocks. *Biofuel Bioprod. Biorefin.* **2018**. [CrossRef]

20. Pradhan, P.; Arora, A.; Mahajani, S.M. Pilot scale evaluation of fuel pellets production from garden waste biomass. *Energy Sustain. Dev.* **2018**, *43*, 1–14. [CrossRef]

21. Tumuluru, J.S. Specific energy consumption and quality of wood pellets produced using high-moisture lodgepole pine grind in a flat die pellet mill. *Chem. Eng. Res. Des.* **2016**, *110*, 82–97. [CrossRef]

22. Tumuluru, J.S. Effect of process variables on the density and durability of the pellets made from high moisture corn stover. *Biosyst. Eng.* **2014**, *119*, 44–57. [CrossRef]

23. Tumuluru, J.S. High moisture corn stover pelleting in a flat die pellet mill fitted with a 6 mm die: Physical properties and specific energy consumption. *Energy Sci. Eng.* **2015**, *3*, 327–341. [CrossRef]

24. Searcy, E.; Lamers, P.; Hansen, J.; Jacobson, J.; Hess, R.; Webb, E. *Advanced Feedstock Supply System Validation Workshop—Summary Report*; Report INL/EXT-10-18930; Idaho National Laboratory: Idaho Falls, ID, USA, 2015; Available online: https://bioenergykdf.net/system/files/1/15-50315-R3_Summary_Report_Only_ONLINE.PDF (accessed on 30 January 2019).

25. Lamers, P.; Roni, M.S.; Tumuluru, J.S.; Jacobson, J.J.; Cafferty, K.G.; Hansen, J.K.; Kenney, K.; Teymouri, F.; Bals, B. Techno-economic analysis of decentralized biomass processing depots. *Bioresour. Technol.* **2015**, *194*, 205–213. [CrossRef] [PubMed]

26. Tumuluru, J.S.; Cafferty, K.G.; Kenney, K.L. Techno-economic analysis of conventional, high moisture pelletization, and briquetting process. Paper No. 141911360. In Proceedings of the American Society of Agricultural and Biological Engineer Annual Meeting, Montreal, QC, Canada, 13–16 July 2014.

27. Pirraglia, A.; Gonzalez, R.; Saloni, D. Techno-economical analysis of wood pellets production from U.S. manufacturers. *BioResources* **2010**, *5*, 2374–2390.

28. Sakkampang, C.; Wongwuttanasatian, T. Study of ratio of energy consumption and gained energy during briquetting process for glycerin-biomass briquette fuel. *Fuel* **2014**, *115*, 186–189. [CrossRef]

29. Granström, K. Emissions of Volatile Organic Compounds from Wood. Ph.D. Thesis, Department of Environmental and Energy Systems, Karlstad University, Karlstad, Sweden, 2005.

30. Johansson, A.; Rasmuson, A. The release of monoterpenes during convective drying of wood chips. *Dry. Technol.* **1998**, *16*, 1395–1428. [CrossRef]

31. Stelte, W.; Holm, J.K.; Sanadi, A.R.; Barsberg, S.; Ahrenfeldt, J.; Henriksen, U.B. Fuel pellets from biomass: The importance of the pelletizing pressure and its dependency on the processing conditions. *Fuel* **2011**, *90*, 3285–3290. [CrossRef]

32. Holm, J.K.; Henriksen, U.B.; Hustad, J.E.; Sorensen, L.H. Toward an understanding of controlling parameters in softwood and hardwood pellets production. *Energy Fuel* **2006**, *20*, 2686–2694. [CrossRef]
33. Van Dam, J.E.G.; Van Den Oever, M.J.A.; Teunissen, W.; Keijsers, E.R.P.; Peralta, A.G. Process for production of high density/high performance binderless boards from whole coconut husk. Part 1: Lignin as intrinsic thermosetting binder resin. *Ind. Crop. Prod.* **2004**, *19*, 207–216. [CrossRef]
34. Serrano, C.; Monedero, E.; Lapuerta, M.; Portero, H. Effect of moisture content, particle size, and pine addition on quality parameters of barley straw pellets. *Fuel Process. Technol.* **2011**, *92*, 699–706. [CrossRef]
35. Mani, S.; Tabil, L.G.; Sokhansanj, S. Effects of compressive force, particle size and moisture content on mechanical properties of biomass pellets from grasses. *Biomass Bioenergy* **2006**, *30*, 648–654. [CrossRef]
36. Shaw, M.D.; Karunakaran, C.; Tabil, L.G. Physicochemical characteristics of densified untreated and steam exploded poplar wood and wheat straw grinds. *Biosyst. Eng.* **2009**, *103*, 198–207. [CrossRef]
37. Carone, M.T.; Pantaleo, A.; Pellerano, A. Influence of process parameters and biomass characteristics on the durability of pellets from the pruning residues of *Olea europaea* L. *Biomass Bioenergy* **2011**, *35*, 402–410. [CrossRef]
38. Puig-Arnavat, M.; Shang, L.; Sarossy, Z.; Ahrenfeldt, J.; Henriksen, U.B. From a single pellet press to a bench scale pellet mill—Pelletizing six different biomass feedstocks. *Fuel Process. Technol.* **2016**, *142*, 27–33. [CrossRef]
39. Jackson, J.; Turner, A.; Mark, T.; Montross, M. Densification of biomass using a pilot scale flat ring roller pellet mill. *Fuel Process. Technol.* **2016**, *148*, 43–49. [CrossRef]
40. Tumuluru, J.S.; Conner, C.C.; Hoover, A.N. Method to produce durable pellets at lower energy consumption using high moisture corn stover and a corn starch binder in a flat die pellet mill. *J. Vis. Exp.* **2016**, *112*, e54092. [CrossRef]
41. Bonner, I.J.; Thompson, D.N.; Plummer, M.; Dee, M.; Tumuluru, J.S.; Pace, D.; Teymouri, F.; Campbell, T.; Bals, B. Impact of ammonia fiber expansion (AFEX) pretreatment on energy consumption during drying, grinding, and pelletization of corn stover. *Dry. Technol.* **2016**, *34*, 1319–1329. [CrossRef]
42. Hoover, A.N.; Tumuluru, J.S.; Teymouri, F.; Moore, M.; Gresham, G. Effect of pelleting process variables on physical properties and sugar yields of ammonia fiber expansion (AFEX) pretreated corn stover. *Bioresour. Technol.* **2014**, *164*, 128–135. [CrossRef] [PubMed]
43. Zafari, A.; Kianmehr, M.H. Factors affecting mechanical properties of biomass pellet from compost. *Environ. Technol.* **2014**, *35*, 478–486. [CrossRef] [PubMed]
44. Tumuluru, J.S. Effect of pellet die diameter on density and durability of pellets made from high moisture woody and herbaceous biomass. *Carbon Res. Convers.* **2018**, *1*, 44–54. [CrossRef]
45. Harun, N.Y.; Afzal, M.T. Chemical and mechanical properties of pellets made from agricultural and woody biomass blends. *Trans. ASABE* **2015**, *58*, 921–930.
46. American Society of Agricultural and Biological Engineers (ASABE). *Cubes, Pellets, and Crumbles—Definitions and Methods for Determining Density, Durability, and Moisture Content*; Standards, S269.4; ASABE: St. Joseph, MI, USA, 2007.
47. *STATISTICA (Data Analysis Software System)*, Version 9.1; StatSoft, Inc.: Tulsa, OK, USA, 2010; Available online: www.statsoft.com (accessed on 18 March 2019).
48. Tumuluru, J.S.; McCulloch, R. Application of hybrid genetic algorithm routine in optimizing food and bioengineering processes. *Foods* **2016**, *5*, 76. [CrossRef] [PubMed]
49. Sahoo, K.; Mani, S. Engineering economics of cotton stalk supply logistics systems for bioenergy applications. *Trans. ASABE* **2016**, *59*, 737–747.
50. Sahoo, K.; Mani, S. Techno-economic assessment of biomass bales storage systems for a large-scale biorefinery. *Biofuel Bioprod. Biorefin.* **2017**, *11*, 417–429. [CrossRef]
51. Tumuluru, J.S.; Lim, C.J.; Bi, X.T.; Kuang, X.; Melin, S.; Yazdanpanah, F.; Sokhansanj, S. Analysis on storage off-gas emissions from woody, herbaceous, and torrefied biomass. *Energies* **2015**, *8*, 1745–1759. [CrossRef]
52. Said, N.; Abdel Daiem, M.M.; Garcia-Maraver, A.; Zamorano, M. Influence of densification parameters on quality properties of rice straw pellets. *Fuel Process. Technol.* **2015**, *138*, 56–64. [CrossRef]
53. Rhén, C.; Gref, R.; Sjöström, M.; Wästerlund, I. Effects of raw material moisture content: Densification pressure and temperature on some properties of *Norway spruce* pellets. *Fuel Process. Technol.* **2005**, *87*, 11–16. [CrossRef]

54. Sarkar, M.; Kumar, A.; Tumuluru, J.S.; Patil, K.N.; Bellmer, D.D. Gasification performance of switchgrass pretreated with torrefaction and densification. *Appl. Energy* **2014**, *127*, 194–201. [CrossRef]
55. MacBain, R. *Pelleting Animal Feed*; American Feed Manufacturers Association: Arlington, VA, USA, 1966.
56. Payne, J.D. Improving quality of pellet feeds. *Milling Feed Fertil.* **1978**, *161*, 34–41.

Article

Characterization of Pure and Blended Pellets Made from Norway Spruce and Pea Starch: A Comparative Study of Bonding Mechanism Relevant to Quality

Anthony Ike Anukam [1,*], Jonas Berghel [1], Stefan Frodeson [1], Elizabeth Bosede Famewo [2] and Pardon Nyamukamba [3]

[1] Environmental and Energy Systems, Department of Engineering and Chemical Sciences, Karlstad University, 651 88 Karlstad, Sweden; jonas.berghel@kau.se (J.B.); stefan.frodeson@kau.se (S.F.)

[2] Electron Microscopy Unit, Central Analytical Laboratory, Faculty of Science and Agriculture, University of Fort Hare, Alice 5700, South Africa; ritayomidef@gmail.com

[3] Technology Station Clothing and Textiles, Cape Peninsula University of Technology, Symphony Way, Bellville 7580, South Africa; nyamukambap@cput.ac.za

* Correspondence: anthony.anukam@kau.se; Tel.: +46-54-700-1664

Received: 19 September 2019; Accepted: 17 November 2019; Published: 20 November 2019

Abstract: The mechanism of bonding in biomass pellets is such a complex event to comprehend, as the nature of the bonds formed between combining particles and their relevance to pellet quality are not completely understood. In this study, pure and blended biomass pellets made from Norway spruce and pea starch were characterized using advanced analytical instruments able to provide information beyond what is visible to the human eye, with intent to investigate differences in bonding mechanism relevant to quality. The results, which were comprehensively interpreted from a structural chemistry perspective, indicated that, at a molecular level, the major disparity in bonding mechanism between particles of the pellets and the quality of the pellets, defined in terms of strength and burning efficiency, were determined by variation in the concentration of polar functional groups emanating from the major organic and elemental components of the pellets, as well as the strength of the bonds between atoms of these groups. Microscopic-level analysis, which did not provide any clear morphological features that could be linked to incongruity in quality, showed fracture surfaces of the pellets and patterns of surface roughness, as well as the mode of interconnectivity of particles, which were evidence of the production of pellets with dissimilarities in particle bonding mechanism and visual appearance.

Keywords: pelleting; functional groups; biomass; pellet strength; combustion efficiency

---

1. Introduction

The use of biomass for the production of energy and valuable chemicals is gaining attention because it is renewable, clean, cheap, and readily available. There are several types of biomass (such as wood, starch, agricultural residues, energy crops, and industrial and municipal solid wastes) suitable for energy conversion after subjecting the biomass to pretreatment processes like pelleting, and the pretreatment method is often determined by the conversion pathway intended for the biomass; however, the need to press biomass into pellets arises from its heterogeneous nature and low energy density, which makes its conversion in energy production systems very problematic [1]. This means that pelleting of biomass is basically a technique used to improve biomass characteristics in the form of a pellet with regular shape, along with higher density, strength, and durability, as well as excellent combustion characteristics and low ash content, which are factors used to define good-quality biomass pellets [2,3]. The improved quality of pelletized biomass makes it suitable as a fuel for use in household heating boiler systems, cooking, and electricity generation. Most biomass pellets are usually made from wood such as Norway spruce; the global fuel pellet market experienced expeditious growth in recent times, and it is anticipated to have even faster growth in the near future [4]. Even though almost all biomass can be pelletized, not all are likely to form durable pellets because of variations in characteristics; hence, different types of biomass are sometimes blended for the purpose of improving quality. Additionally, because of issues related to dust formation and self-ignition, additives such as starch may be added to the biomass during pelleting in order to increase the overall quality of the pellets [2,5]. However, the production of durable biomass pellets is always challenged by a host of factors, including a lack of fundamental understanding of the bonding mechanism of major components during pelleting, type of materials to be blended with the biomass, how these materials affect the mechanism of bonding, and different pellet quality parameters. Other factors impacting the production of durable biomass pellets include types of biomass for pelleting, moisture content of the biomass, organic and elemental constituents of the biomass, particle size and distribution, and pellet press compression force and temperature [6–9]. Most of these factors were studied by other researchers, yet differences in the mechanism of bonding between pure and blended biomass pellets relevant to quality still need to be investigated. As previously mentioned, wood is a type of biomass most commonly used in the production of fuel pellets and remains a major source of energy in most countries. On the other hand, starch is perceived as a good additive to wood for the production of durable biomass pellets. However, to the best of the authors' knowledge, the production of biomass pellets from pure starch and other materials (such as wood) blended with starch, with the aim to investigate the mechanism of bonding relevant to quality, is sparsely studied. To lay the groundwork for a better understanding of what was investigated in this study, the section below presents a brief synopsis of the chemistry of wood and starch.

1.1. Overview of the Chemistry of Wood and Starch

It was reported that wood such as Norway spruce remains the most common material for the production of fuel pellets in Sweden, and that starch has excellent applications in a handful of industrial sectors such as biofuel, food, pulp, and paper [5,10]. The properties of these two materials are controlled by complex interactions between their physical and chemical structure. For instance, the structural characteristics of wood are such that its cells are made up of varying proportions of three major substances (cellulose, hemicellulose, and lignin), whose structures are bound by functional groups that are responsible for the behavior of wood in many conversion processes. Depending on the type of wood, the weight percentages of cellulose, hemicellulose, and lignin in wood range from 40% and 50% for cellulose, from 20% to 28% for hemicellulose, and from 25% to 30% for lignin [3,7,11,12]. The wood structure is complex and anisotropic with optimized hierarchical levels that span from the macro to the micro, molecular, and even nano scale. The structure of the cellulose component of wood is linked by β-(1,4)-glycosidic bonds with a high degree of polymerization, while that of hemicellulose is partially substituted by acetyl groups with a lower degree of polymerization in comparison to cellulose. The substituted acetyl groups in the hemicellulose structure means that the hydroxyl groups (–OH) at carbon positions C_2 and C_3 are partly substituted by *O*-acetyl groups, one of the adhesive degradation products of hemicellulose responsible for natural bonding [13,14]. Lignin is complex in nature and contains aromatic rings that are responsible for its glue effects; the thermosetting properties of lignin are exhibited at temperatures above or equal to 100 °C, and the adhesive nature of thermally melted lignin significantly contributes to the strength and durability of pellets made from lignocellulosic biomass [15–17].

Starch, on the other hand, is a soft, white, and tasteless powder whose usefulness as an additive in improving bonding properties of biomass pellets cannot be overemphasized. This is because of the chemical structure of its two major monomer units known as amylose and amylopectin [18]. Just like wood, the structures of these two monomer units of starch are also bound by functional groups that confer specific properties. The functionality of starch depends on these groups and the average molecular weight of its amylose and amylopectin contents [19]. Both of these constituents of starch consist of chains of α-(1,4)-glycosidic bonds that are linked by D-glucose residues connected via α-(1,6)-glycosidic linkages, thereby forming polymer branches [19,20]. More often than not, the relative weight percentages of amylose and amylopectin in starch range from 18% to 33% for amylose, and from 72% to 82% for amylopectin [19]. Morphological features of starch include perfectly spherical particles with many void spaces exhibiting both internal and external surface areas that are determined by the shape and size of its particles, which induces higher molecular mobility when starch is heated (gelatinization) [19,21].

The structures of the major components of both wood and starch can be found in References [12,22].

1.2. Purpose of Study

Several studies reported the pelleting of different types of biomass and their blends, with most of the studies focusing on material and process parameters including the mechanism of bonding from different viewpoints [2,23–27]. Despite the numerous studies, the exact mechanism involved in bonding in biomass pellets and the source of inter-particle bonding relevant to pellet quality are not fully understood and still a subject of debate, particularly when the pellets are made from a blend of two or more different biomass materials. Not much was undertaken, in terms of research, to study and establish, from a structural chemistry perspective, differences in the mechanism of bonding between pure and blended biomass pellets relevant to the production of good-quality pellets. Investigations of this sort require a systematic approach that takes into account the use of state-of-the-art analytical techniques that can go beyond standard tests and visual assessments. The techniques must be able to provide information beyond what can be visualized by the human eye and offer evidence of quality without having to rely on special quality assessment tools. Therefore, this study aims to investigate and establish differences in the mechanism of bonding, relevant to quality, between pure and blended pellets made from Norway spruce and pea starch, relying on molecular, microscopic, and thermal analyses provided through the use of advanced analytical instruments able to offer information beyond what can be visualized by the naked eye.

The choice of the materials used in the study, the reason for blending, and the blend ratio are explicitly described in the section below.

2. Materials and Methods

2.1. Sample Preparation Steps

While Norway spruce was selected for this study because of its availability and suitability as a feedstock for heat and power production, pea starch was chosen for its cost-effectiveness and suitability as an additive in pelleting processes. Whereas the pea starch was an industrial secondary starch obtained in powder form from Lyckeby, Kristianstad, Sweden, Norway spruce was obtained in the form of logs from a local sawmill in Karlstad, Sweden and was cut into chips. The two raw materials, upon procurement in their pure form, were placed in small plastic bags and stored at room temperature. Prior to pelleting, the materials were dried in an oven at 50 °C for 24 h. Because starch was obtained in powder form, it did not require milling, but was sieved to a particle size of 0.3 mm, according to American Society for Testing and Materials (ASTM) standards (D 2013-72). Norway spruce was milled to a particle size of <1 mm using a Wiley mill and also sieved to obtain a final particle size equal to that of pea starch. Appropriate quantities of each sample were separately weighed and mixed with water to obtain a final moisture content of 10% (wet basis). This was because moisture is often required during pelleting to reduce friction and to avoid blocking of the pellet press die; under these circumstances, however, adding the right amount of water is very vital, as material moisture content must not exceed 10% for pellets made from woody biomass, and a maximum moisture content of 15% for pellets made from grass [6,28]. Preceding these steps and soon after procurement, milling, and sieving, the initial moisture contents of the two samples (Norway spruce and pea starch) were determined according to standards [29]. The determination of moisture content was done on a wet basis by weighing the two samples and placing them in an oven at 105 °C for 24 h and weighing again to determine final weight.

2.2. Preparation of Blend

A 50/50 blend ratio of milled Norway spruce and pea starch was prepared by weighing 45 g of each respective material in order to achieve a uniform weight ratio for blending. Both materials were placed in a plastic container and thoroughly mixed together for 10 min using a 220-V Janke and Kunkel Ika-Werk mixer rotated at a speed of 60 rpm. The blend was prepared by mixing with water to achieve a moisture content (10%) equal to that of the pure samples described in the previous section. The procedure for achieving the desired moisture content for the blend was such that an appropriate amount of water was added to 90 g of the mixture and then mechanically rotated again for another 10 min. Shortly after rotation, the mixture was sealed in small plastic bags and stored for 48 h to allow the moisture to settle in and to prevent further absorption of moisture from the surroundings. Care was taken during the preparation of mixtures with water so as to avoid the formation of rigid irreversible gel by pea starch, which may alter the analysis results. The 50/50 percentage blend was a random selection aimed at keeping the ratio of the admixture of Norway spruce and pea starch equal for a fair comparison of the mechanism of bonding between the pure and blended pellet samples made from the two materials. The pellet production process of the pure and blended samples of Norway spruce and pea starch is described in the section below.

2.3. The Pellet Production Process

Pure and blended pellets were produced from Norway spruce and pea starch using a single pellet press available at the Department of Engineering and Chemical Sciences of Karlstad University in Sweden. Figure 1 presents the laboratory-scale pellet press set-up used in the pellet production process.

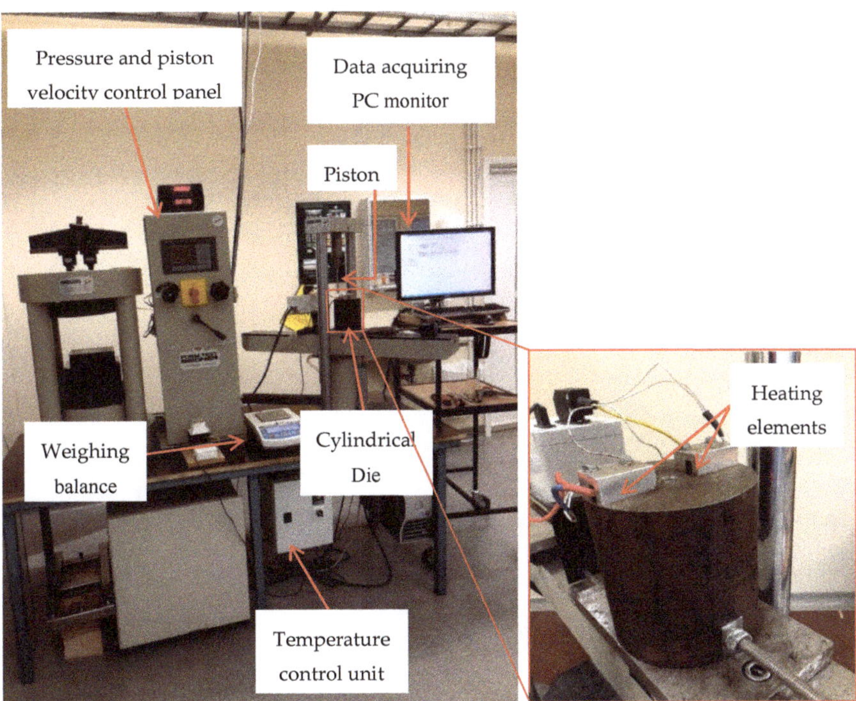

Figure 1. The single-unit pellet press set-up used in this study with labeled parts.

The press consisted of a cylindrical die made of hard steel, about 8 mm in diameter, with connected heating elements and thermal insulation. A temperature control unit was used to regulate die temperature from ambient conditions to a maximum temperature of 100 °C. The bottom end of the cylindrical die was closed with a removable plastic piston stopper. Compression force was mechanically applied with a piston pointing directly toward the top opening of the die filled with exactly 1 g each of the pure and blended samples of milled Norway spruce and pea starch. Piston movement was controlled by the entire pellet press system, and pressure was applied to the piston through a load cell whose maximum compression force was about 2000 kN. Before each use and when changing the raw materials, the die was rinsed several times with acetone and cleanly wiped with a paper towel. Each of the pure and blended samples was compressed at a velocity of 30 mm·min^{-1} until a maximum pressure of 16 kN was reached. The holding time for every test was set at 10 s, which is in agreement with studies conducted by Nielsen et al. [30] and by Nguyen et al. [31]. The pellets were removed from the die by removing the stopper and mechanically pushing out the pellets, slowly, at a velocity of 5 mm·min^{-1}. According to a previous study [7], the optimum pelleting conditions under which the pure and blended samples of Norway spruce pellet (NSP) and pea starch pellet (PSP) were produced at a laboratory scale are shown in Table 1.

Table 1. Optimum pelleting conditions under which pure and blended samples of Norway spruce pellet (NSP) and pea starch pellet (PSP) were produced [7].

Parameter	Pelleting Conditions
Moisture content	10% (prior to pelleting)
Die temperature	100 °C
Compression force	16 kN
Holding time	10 s
Piston velocity	30 mm·min^{-1} (compression phase)
Push out velocity	5 mm·min^{-1} (friction phase)
Raw material quantities	1 g (each of milled material)

2.4. Analytical Techniques

It is worth mentioning that, in addition to the data generated on differences in the bonding mechanism of the pure and blended samples of NSP and PSP used in this study, the quality of the pellets was interpreted in terms of strength and combustion efficiency. The study did not incorporate the use of special tools that are restricted to the determination of the quality parameters; instead, the study relied on evidence of the above quality parameters from the data obtained through the use of state-of-the-art molecular, microscopic, and thermal analytical instruments. Therefore, the subsections below present specific information about the analytical instruments used in this study and describe the experimental procedures related to how the instruments were used. For the sake of reproducibility, all analyses presented in this study were repeated in triplicate, and average values were reported.

2.4.1. Content Analysis

Assessing biomass for any conversion or pretreatment process often requires a good understanding of its basic composition and characteristics [1,3,32]. Thus, the major components of the pure and blended pellet samples believed to have greater influence on bonding and pellet quality were considered during the content analysis. Content analysis of both pure and blended samples of NSP and PSP was conducted to determine major organic and elemental components in terms of percentage composition. Details of how these components were quantified are given in Section 2.4.3. However, subsequent analytical instruments were used to investigate and establish differences in the mechanism of bonding between the pure and blended pellet samples relevant to bonding and quality of the pellets.

The major organic components of PSP (100%) were quantified by a differential scanning calorimeter (DSC), whose make, model, and principle of operation are detailed in Section 2.4.4.

For the major elemental constituents of the pure and blended pellet samples, an energy-dispersive X-ray spectroscope (EDX) connected to a scanning electron microscope (SEM) was used. The analysis (EDX) involved the generation of characteristic X-rays that revealed elements present in the pellet samples. The procedure was such that, the SEM, whose model and mode of operation are detailed in Section 2.4.5, created an image of the samples by scanning them with an electron beam. During this process, an X-ray was emitted from the samples, which was then collected by the EDX instrument after bombardment by an electron beam. The collected X-rays were characteristic of the atomic structure of the elements contained in the pellet samples. Energy level was used to sort the X-rays, after which a plot of X-ray energy versus frequency was obtained. This plot gave an indication of the elements present in the pellet samples and the percentage composition of each element.

The results obtained from this analysis (content analysis) laid the groundwork for data interpretation from subsequent analysis involving the use of other advanced analytical instruments.

2.4.2. Fourier-Transform Infrared Spectroscopic Analysis (FT-IR)

Since bonding, from a chemistry perspective, relates to perpetual attraction between atoms, ions, or molecules, there is a need to firstly identify the type of forces acting between individual particles of the pure and blended samples of NSP and PSP prior to identifying chemical alterations caused by blending. However, the goal of this analysis was to determine the influence of chemical alteration on bonding and establish differences in the mechanism of bonding between the pure and blended pellet samples relevant to the quality of the pellets. Analysis of this nature would require molecular-level identification of key functional groups with significant impact in bonding. Therefore, FT-IR analysis was invaluable in helping to achieve this goal. The procedure for the FT-IR analysis is briefly described below.

FT-IR measurements of pure and blended samples of NSP and PSP were carried out using a Perkin Elmer FT-IR spectrometer (Spectrum Two). Spectra were scanned at a resolution of 4 cm^{-1} with 32 scans collected in the spectral range of 4 000 cm^{-1} to 500 cm^{-1}.

2.4.3. Thermogravimetric Analysis (TGA)

Combusting biomass at elevated temperatures creates chemical modifications that affect the properties and performance of the biomass in any thermal pretreatment or conversion process, and the extent of chemical modification depends on the temperature level and the duration of conditions of thermal exposure [33,34]. In thermal analysis of biomass materials, physical and chemical changes associated with the material are usually determined as a function of temperature. TGA analysis of pure and blended samples of NSP and PSP was conducted for the following reasons: (1) to determine and quantify major organic constituents based on the degradation temperature of each component; (2) to determine differences in the temperature range of chemical modifications relevant to bonding and pellet quality; (3) to determine the combustion characteristics/efficiency of the pellets relevant to quality. Since the pellet samples were made from two different materials (Norway spruce and pea starch), TGA analysis was undertaken to specifically establish how, if any, differences in modification temperature can facilitate particle bonding. A plot of weight loss against temperature is usually obtained from TGA analysis and represents the thermal behavior of a sample and the media through which components of the sample can be quantified based on the degradation temperature of each component [7,35]. Therefore, the major organic components of NSP (100%) and the NSP/PSP (50%/50%) blend were quantified from the plot obtained from this analysis, done in accordance with (after moisture evaporation) differences in the thermal decomposition temperatures of each component, as the pellet samples lost weight as temperature increased. The major organic components of biomass are commonly differentiated and identified using TGA due to differences in the structure of the components [33,36–38]. The procedure for TGA analysis is described below.

Thermal analyses of pure and blended samples of NSP and PSP were performed with a Perkin Elmer TGA 4000 instrument in which samples were firstly placed in the combustion chamber of the instrument and ignited via software programming. The samples were heated from about 28 to 760 °C at a heating rate of 20 °C·min^{-1} under a nitrogen gas flow rate of 19.8 mL·min^{-1} in order to maintain a non-reactive atmosphere as the analysis progressed. The weight loss of the samples was typically monitored as the samples were heated, cooled, and isothermally held.

2.4.4. Differential Scanning Calorimetric Analysis (DSC)

According to Back [39], sufficient bonding areas are activated when biomass components are plasticized above their transition temperatures (T_g). The comparative assessment of combustion profiles helps to determine the series of stages that characterize the thermal behavior of biomass [40]. DSC analysis was used as a complementary technique to TGA to determine the T_g or softening temperature of the pellet samples as a function of heat flow, as well as establish phase changes caused by thermal exposure as result of blending. The intention was to compare the T_g of the pure and blended pellet samples and determine if this would have an impact on how bonding occurs in pelleting of different biomass materials. Polymer viscosity significantly deceases to show obvious flow characteristics that are relevant to bonding at T_g higher than 50 °C [7,41]. As previously mentioned, the major organic and elemental components of PSP (100%) used in this study were determined by DSC according to the method described by Moorthy et al. [42].

A Perkin Elmer DSC 6000 was used to determine T_g caused by thermal degradation at specific heat flow during thermal analysis of pure and blended samples of NSP and PSP. The samples were heated to a temperature of approximately 440 °C at a heating and nitrogen gas flow rates similar to those given for TGA analysis. A computer was used to monitor the temperature and regulate the rate at which temperature changed for a given amount of heat.

2.4.5. Scanning Electron Microscopic Analysis (SEM)

To understand sample composition and morphological characteristics relevant to bonding at a micro scale, as well as view structural changes that are vital factors of pellets quality in terms of strength, a high-resolution SEM that offers surface morphology of samples at high magnifications was used. The intention was to use images generated by the SEM to determine visible differences in particle bonding mechanism between the pure and blended pellet samples that were relevant to quality through the identification of quality features from the images.

The morphological characterization of pure and blended samples of NSP and PSP was conducted with a JEOL (JSM-6390LV) model SEM instrument fitted with an EDX analyzer, which was used for the analysis of major elemental components. Samples were placed on an aluminum holder stub using a double-sided sticky carbon tape and sputter-coated with gold (Au) using an Eiko IB3 Ion Coater. Sputter-coating was necessary to increase conductivity and prevent charging effects that may lead to blurred images, which may hamper detailed interpretation of results. Thereafter, samples were mounted on stub holders in the sample chamber of the SEM for morphological examination.

3. Results

3.1. Compositional Analysis

The results obtained from this entire study were comprehensively interpreted from a structural chemistry viewpoint and were also used to predict the relevance of bonding to the quality of the pellets. The presented data from the analytical instruments used in the study were substantiated using information from the existing literature.

As previously mentioned, compositional analysis was undertaken to determine major organic and elemental components as a first step to the study. Table 2 presents the percentage composition of the major organic and elemental components of pure and blended samples of NSP and PSP obtained from the content analysis test presented in Section 2.4. The percentage error for the reported values was within ±0.3 and 1.0% for the major organic constituents, and within ±0.7 and 1.2% for the main elemental components.

Table 2. Percentage composition of major organic and elemental components of pure and blended samples of NSP and PSP. ND—not determined.

	Major Organic Constituents (%)		
	NSP (100%)	**PSP (100%)**	**NSP/PSP (50%/50%)**
Cellulose	41.6	ND	21.7
Hemicellulos	23.2	ND	12.3
Lignin	32.9	ND	16.5
Amylose	ND	28.2	19.2
Amylopectin	ND	70.3	27.8
	Major Elemental Components (%)		
C	46.2	33.4	40.8
H	6.4	5.8	5.2
O	44.6	58.3	51.3

It should be noted that NSP (100%), PSP (100%), NSP/PSP (50%/50%), and pure and blended pellet samples were used interchangeably throughout the article.

Since content analysis for this study focused on determining the percentage composition of major organic and elemental constituents of the pure and blended pellet samples, the data in Table 2 were interpreted in relation to variation in composition.

From Table 2, it can be noted that the major organic components of NSP (100%) differed from those of PSP (100%), indicating that woody biomass (like Norway spruce) is made up of different organic constituents in comparison to pea starch, whose organic constituents are composed of varying proportions of amylose and amylopectin. These two components (amylose and amylopectin) suggest that PSP (100%) is equally polysaccharide in nature. The balance of mechanical stability and durability offered by starch when blended with other materials in a pelleting process arises from its polysaccharide nature as a result of the chemical structures of its two major macromolecular components (amylose and amylopectin) [18]. In contrast to the pure pellet samples (100% each of NSP and PSP), the blend (NSP/PSP 50%/50%) showed a significant decrease in its organic constituents, a condition ascribed to component consolidation and the nature of the blended materials. This was basically viewed as a change in composition and structure that was construed to mean an alteration in chemical characteristics and an increase in strong intermolecular chain interactions. When starch is blended with materials such as wood, a product with modified properties and structure is obtained; the modification is facilitated by intermolecular chain interaction [43]. The alteration in the chemical characteristics of the blend (NSP/PSP 50%/50%) played a significant role in accelerating the tendency for particle-to-particle bonding because major organic components were consolidated upon blending. This was proven by complementary analytical techniques whose data are presented in the sections below. Nonetheless, the significant reduction in the major organic components of the blend also implied a reduction in the concentration of active bonding groups, a fact established by the FT-IR data presented in the next section.

For the elemental components of the pellet samples, the data in Table 2 indicate that C, H, and O constituted major elemental components of the pure and blended pellet samples of NSP and PSP used in this study. However, significant variations in the percentage composition of these elements could be noted, and the reason for the variation is the same as that given for the content of major organic components of the pellets.

It is noteworthy to mention that the percentage composition of the major organic and elemental components of the pure and blended pellet samples of NSP and PSP presented in Table 2 did not add up to 100% because the left-over fractions were ceded to minor components (such as extractives and proteins) with percentage composition <3% and were considered insignificant; hence, they were not reported.

3.2. FT-IR Analysis

In addition to the goals of undertaking FT-IR analysis described in Section 2, data from the analysis also helped compare and establish differences in the mechanism of bonding between the pure and blended pellet samples relevant to quality. The FT-IR spectra of pure and blended pellet samples of NSP and PSP are presented in Figure 2. The data were obtained from the experimental analysis described in Section 2.4.2. To permit convenient comparisons, the spectra are presented as a single plot with the legends clearly displayed in the plot and spectral bands numbered for clarity.

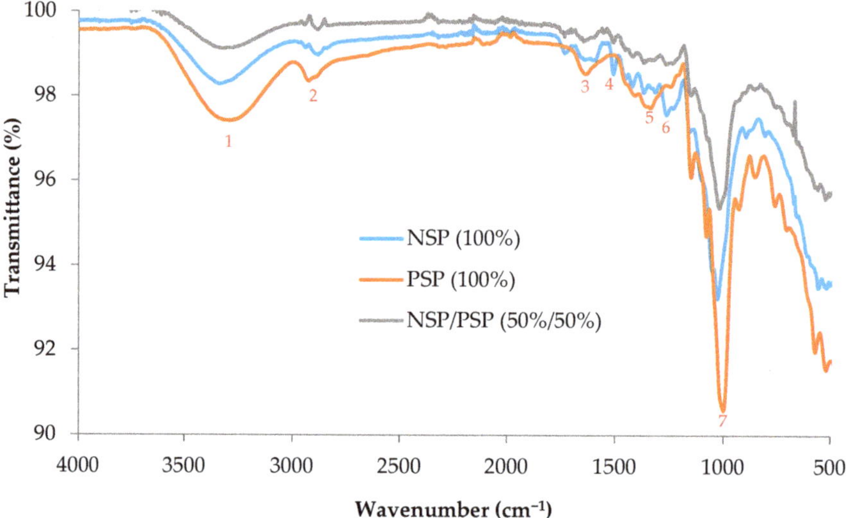

Figure 2. Fourier-transform infrared (FT-IR) spectra of pure and blended samples of Norway spruce pellet (NSP) and pea starch pellet (PSP).

Prior to interpretation of the FT-IR spectra in Figure 2, it is vital to mention that some spectral bands are more relevant than the others in the differentiation of the content (in terms of functional groups) of the pure and blended pellet samples. To avoid ambiguity, numbers were used to represent the most important bands considered relevant to the type of attraction forces holding particles of the pellets together, and to the quality of the pellets.

It is obvious from Figure 2 that the spectra were quite similar to each other, which does not necessarily mean the same characteristics. Using the pure pellet samples (100% each of NSP and PSP) as reference points, it is also apparent that the composite material (NSP/PSP 50%/50%) showed a synergistic effect. This synergistic behavior was more pronounced around 2893 to 1254 cm^{-1} (bands 2–7), which were bands associated with active bonding groups such as C=O, C–O, C–O–C, and C–H, attributed to amylose and amylopectin, as well as cellulose, hemicellulose, and lignin in all three pellet samples [7]. All bands in this region (bands 2–7) had different intensities in the spectra of the three pellet samples. Some synergy can also be observed in band 2 of the spectra. The wide transmittance band around 3312 cm^{-1} (band 1) for all samples was assigned to the hydroxyl group (–OH), an active bonding group responsible for band broadening due to a combination of intra and intermolecular hydrogen bonding. The presence and concentration of active bonding groups determine the type of attraction forces between individual particles of biomass pellets [7]. According to Popescu et al. [44], the broadening of the –OH band in IR spectra is often caused by the presence of intra and intermolecular hydrogen bonds. However, the –OH group region of the IR spectrum is particularly useful for explaining patterns of hydrogen bonding, because each distinct –OH group offers a single stretching at a frequency that decreases with increasing strength of the hydrogen bonds, which are responsible for various properties of cellulose, hemicellulose, and lignin that are associated with NSP (100%) and NSP/PSP (50%/50%) [44].

Band intensities of samples in FT-IR analyses are true representations of the concentration of components [45]. Therefore, the active bonding groups previously listed absorbed light with greater intensity and indicated the presence of polysaccharides such as starch [46], leading to the more pronounced band intensity of PSP (100%). This suggests that PSP (100%) contained higher proportions of polysaccharides with greater amounts of active bonding groups (C–O and –OH groups) capable of stronger dipole–dipole attraction and hydrogen bonding (which are two types of intermolecular forces of attraction) as compared to pure NSP (100%) and NSP/PSP (50%/50%). Nonetheless, in the spectrum of NSP (100%), the transmittance band around 1275 and 1478 cm^{-1} (bands 4 and 6) was due to distance-dependent interactions associated with van der Waals contact distances as a result of groups such as C=O and C–H, which are known to form bands within the wavelengths specified above. In addition to being active bonding groups, these groups are equally polar in nature because of their ability to form different types of attraction forces [7]. Thus, for the blend (NSP/PSP 50%/50%), it was presumed that its particles were held together by strong hydrogen bonding because of the presence of the C=O group, which is a polar functional group capable of initiating structural changes by introducing more intermolecular forces of attraction.

With respect to the quality of the pellets, defined in this study in terms of strength and burning efficiency, it is fair to allude that, in relation to the strength of the pellets, PSP (100%) seemed to be a higher-quality pellet than NSP (100%) and NSP/PSP (50%/50%). This was because of its more pronounced band intensity, which was construed to mean greater proportions of polar functional groups that were capable of forming a combination of dipole–dipole attraction and hydrogen bonding between particles. A union of these two attraction forces creates stronger bond energies/bond strength, whereas the functional groups associated with NSP (100%) formed more of van der Waals attraction forces than hydrogen bonding between particles of the pellet, which may increase the possibilities of the formation of fines. Particles of the blend (NSP/PSP 50%/50%) were more connected by a combination of multiple forces of adhesion and cohesion, some of which may have reduced bond strengths that were easily broken. These findings are in agreement with the fact that the quality of biomass pellets, among other factors, depends on the type and strength of attraction forces between particles, as alluded by Kaliyan [41]. In view of this, the order of strength of the pellets was as follows: PSP (100%) > NSP/PSP (50%/50%) > NSP (100%). For the quality of the pellets in terms of burning efficiency, the thermal analysis data presented in Sections 3.3 and 3.4 has provide the needed information.

It is worth mentioning that spectral interpretation from FT-IR analysis of samples is non-exhaustive. Thus, for clearer spectral interpretation and understanding, the peak absorption range of different functional groups can be found in Reference [47].

3.3. TGA Analysis

The determination of the modification range temperature, the effects of blending, and the comparative assessment of the mechanism of bonding relevant to pellet quality are among important instances where thermal analysis addresses specific needs. One of the tools proven to address these needs is a thermogravimetric analyzer (TGA). Thus, knowledge of the major compositional changes that take place as biomass undergoes thermal treatment can facilitate understanding of its behavior in pretreatment processes including in pelleting processes [48]. Figure 3 shows the TGA thermograms of the pure and blended pellet samples with the focus area clearly marked for less ambiguous interpretation.

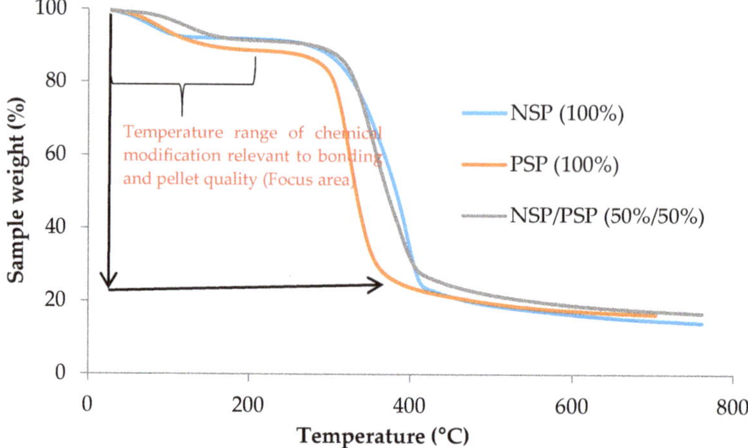

Figure 3. Thermogravimetric analysis (TGA) thermograms of pure and blended samples of NSP and PSP.

Interpretation of the TGA plot in Figure 3 was limited to the focus area marked in the plot. Events beyond the focus area were not considered because, according to some studies [3,5,7,8,41,49,50], the optimum temperature range for pelleting biomass falls between 50 and 150 °C. Within this temperature range, flow and combustion characteristics relevant to bonding and pellet quality are usually formed [5,7,8]. However, it is important to bear in mind that thermal degradation of a sample under TGA proceeds differently in terms of the pattern, based on how the reduction in sample weight occurs as the sample undergoes combustion [51]. This means that the weight change of the sample (increase or decrease in weight) as temperature increases depends on a host of factors, including the type of sample, its stability, and analysis conditions [33,35,52]. Having said this, while the thermal behavior of NSP (100%) was controlled by its major organic components (cellulose, hemicellulose, and lignin), that of PSP (100%) was equally influenced by its primary organic macromolecular constituents (amylose and amylopectin), and the combustion behavior of the blend (NSP/PSP 50%/50%) was influenced by a combination of the contents of both materials (Norway spruce and pea starch).

Eye-tracking TGA curve information from Figure 3 shows that the weight change of all three pellet samples was completed at about 760 °C. The first noticeable feature, however, was the reduction in weight of the samples as temperature increased (indicated by the arrows), a condition which forms the core of the thermal performances of biomass materials [7,33]. Another observable feature is the differences in the degradation patterns of the curve, particularly around the focus area (focus point). This point represents the temperature range of chemical modification in which water molecules and other small molecular species relevant to bonding are liberated [7]. The release of moisture began at almost the same temperature for all samples (around 45 °C), but the temperature at which molecular species began to flow was slightly different between the pure and blended pellet samples, an indication of differences in thermal behavior attributed to compositional disparity. This temperature for NSP (100%) was about 55 °C, around 65 °C for PSP (100%), and ca. 60 °C for NSP/PSP (50%/50%). Shortly after moisture evaporation and the release of molecular species, critical evaluation of the plot showed that the degradation temperature for NSP (100%) relevant to pelleting began at ca. 85 °C, while that for the blend (NSP/PSP 50%/50%) started at about 75 °C, with that for PSP (100%) beginning at a moderately higher temperature of above 90 °C. This typically means that the release of the bonding components of PSP (100%) required more energy than those of NSP (100%) and NSP/PSP (50%/50%), a condition believed to be attributed to higher concentrations of polar functional groups, particularly the C–H group (from band 7 in the FT-IR spectra in Figure 2) in the structure of PSP (100%) with greater bond strength and bond energy as compared to NSP (100%) and NSP/PSP (50%/50%). The bond strength and bond energies of different functional groups, as well as those of various attraction forces, can be found in References [53,54]. Atoms of polar functional groups are held together by a combination of strong dipole–dipole attraction and intermolecular forces that require more energy to be broken, resulting in the thermal behavior of PSP (100%) being different than that of NSP (100%) and NSP/PSP (50%/50%), for which the energy required for bond breaking would be much less. The more concentrated these groups are in a material, the greater is the energy required to break the bonds holding atoms of these groups together, which can also be construed to mean that the stronger a bond is, the higher the energy required is to break the bond [7,50,54]. A review of the literature [53,54] also showed that the bond energies of different functional groups and those of different types of attraction forces vary.

For the quality of the pellets in terms of burning rate and efficiency of combustion, this analysis (TGA) established that NSP (100%) may be a better quality pellet in comparison to PSP (100%) and NSP/PSP (50%/50%) because of its lower modification temperature, which was construed to mean faster burning velocity that translates into better combustion efficiency due to greater contact between the pellet sample (NSP 100%) and the heating environment. Biomass combustion efficiency is a function of good contact between the heating environment and the biomass [55]. According to Gil et al. [56], the burning rate of pure biomass pellets, which is determined by temperature, is usually higher than that of their blends. Furthermore, judging by the FT-IR data of the pellets presented in Figure 2, there were higher percentages of oxygen-containing polar functional groups for PSP (100%). This puts PSP (100%) at a great disadvantage as a feedstock in thermal energy conversion systems due to poor combustion-related issues that may significantly increase the operating and maintenance costs of the energy systems. This is because oxygen-containing polar functional groups like the –OH groups are hydrophilic in nature and have the ability to retain moisture, thereby increasing the viscosity of starch micelles (starch gelatinization) [57,58]. The issues of starch gelatinization are explained in greater detail in a correlative thermal analysis data presented in the section below. In the case of the blend (NSP/PSP 50%/50%), however, the oxygen-containing polar functional groups may have been compromised or compensated for by the cellulose, hemicellulose, and lignin contents of Norway spruce. In view of these results, therefore, the order of quality of the pure and blended samples of NSP and PSP in terms of burning rate and efficiency of combustion was as follows: NSP (100%) > NSP/PSP (50%/50%) > PSP (100%).

It is vital to mention that there are no standard methods to determine combustion efficiency; hence, data from thermal analysis of biomass pellets aimed at burning efficiency cannot be compared without uncertainties [59].

3.4. DSC Analysis

According to Back [39], sufficient bonding areas are created when wood components are plasticized above their glass transition temperature (T_g). In addition, since combustion is the final destination of most biomass pellets, an ultimate method for testing the quality of biomass pellets is thermal analysis, because good-quality pellets are expected to burn easily and generate less complex degradation patterns [60,61]. Figure 4 presents the DSC plot of pure and blended samples of NSP and PSP.

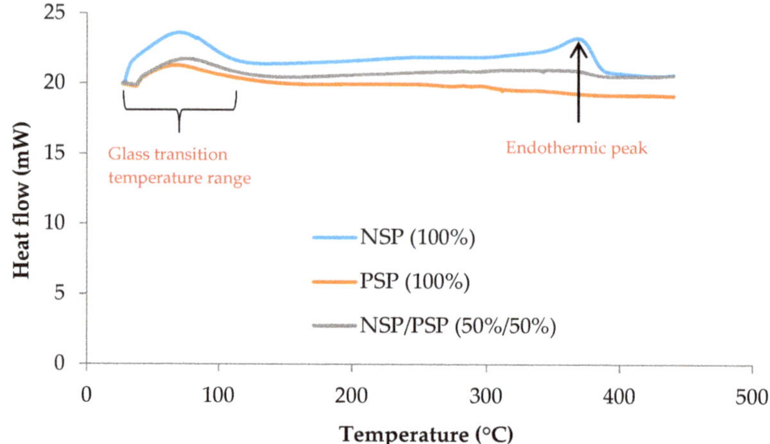

Figure 4. Differential scanning calorimetry (DSC) thermograms of pure and blended samples of NSP and PSP.

A step change in the heat flow of the DSC thermogram presented in Figure 4 constituted the first observable feature of the plot and represented the glass transition (T_g) range temperature for all samples. T_g conditions facilitate the deformation of particles, reduce viscosity, and increase the movement of natural binding components [41]. According to the DSC plot in Figure 4, these temperature conditions fell between 45 and 85 °C for NSP (100%), between 45 and 80 °C for PSP (100%), and between 45 and 75 °C for NSP/PSP (50%/50%). Moisture is one of the most useful agents that can act as a lubricant during pelleting of biomass materials because it strengthens and promotes bonding through a combination of attraction forces by increasing the particle contact area [3,41,62]. Just like the previous thermal analysis data, transition at 45 °C in the DSC plot typically indicated the moisture evaporation from all samples. The T_g range of the samples under DSC analysis were consistent with the previous modification temperature range presented in Section 3.3, except that, in this analysis (DSC analysis), the maximum T_g for PSP (100%) occurred at a much lower temperature (80 °C) than its previous maximum modification temperature of around 90 °C in the previous section. The complexity of the thermal behavior of starch and differing analysis conditions may be responsible for the difference in the maximum modification and transition temperatures between this thermal analysis and the previous for 100% PSP. The behavior of starch under thermal conditions is much more complex than other thermoplastic materials like wood because the physicochemical changes that occur during combustion of starch or products containing starch involve the motion of water, gelatinization, glass transition, and melting, as well as the change of crystal structure and molecular degradation, which are thermal events that are dependent on material moisture content, and the water contained in starch is often not stable during combustion [63].

The DSC plot in Figure 4 also indicated that 100% NSP displayed two endothermic peaks, with its first in the T_g range, and its second way beyond the T_g range (between 350 and 370 °C). However, further interpretation of the DSC plot beyond the Tg range for all samples was less of a focus because the study aimed to establish differences in the mechanism of bonding between the pellet samples relevant to quality, and the T_g is considered the temperature range at which natural binders relevant to bonding are activated [6,41]. Therefore, inter-diffusion of the natural binders in 100% NSP increased between adjacent particles to facilitate bonding and the formation of solid bridges at its T_g. The polymeric constituents of the blend (NSP/PSP 50%/50%) were released faster than those of the pure pellet samples (NSP 100% and PSP 100%) because of lower T_g and the presence of water vapor created by moisture. Differences in T_g between the pure and blended pellet samples can be traced back to their content of polar functional groups (Figure 2) including the oxygen-containing groups whose concentrations were higher with stronger bond energies in PSP (100%) than NSP (100%) and the blend (NSP/PSP 50%/50%).

For the quality of the pellets in terms of combustion efficiency, which was considered in this analysis (DSC analysis) from the viewpoint of burning and heat flow rates, NSP (100%) seemed to dominate because of the two endothermic peaks formed in its DSC plot (Figure 4). The formation of the two endothermic peaks and the modification temperature for NSP (100%) from the previous thermal analysis data (55 °C as compared to 60 °C and 65 °C for NSP/PSP 50%/50% and PSP 100%, respectively) were evidence of increased burning and heat flow rates that were facilitated by auto-oxidation reactions because of the presence of oxygen-containing functional groups in the structure of NSP (100%). This means that the burning and heat production rates of NSP (100%) were significantly higher in comparison to those of PSP (100%) and NSP/PSP (50%/50%). Although, PSP (100%) may have greater concentrations of oxygen-containing polar functional groups, there is the possibility of the formation of hydrogen bridges at elevated temperatures because of increased probabilities of the promotion of retrogradation reactions that may result in the formation of thickened paste, which reduces combustion efficiency. However, this thickened paste can facilitate particle-to-particle bonding to form pellets with increased strength due to the presence of rigid irreversible gel [64]. The concentrations of the oxygen-containing polar functional groups in the blend were compromised by mixing the two materials at a 50/50 ratio. However, because of the balance between the ratio of the two materials in the blend (NSP/PSP 50%/50%), and perhaps the higher percentage of carbon in NSP (100%) in comparison to PSP (100%), combustion issues would be minimized when the blended pellet is used as feedstock in thermal conversion systems. Therefore, the order of quality of the pellets in terms of combustion efficiency was as follows: NSP (100%) > NSP/PSP (50%/50%) > PSP (100%).

3.5. SEM Analysis

When biomass is heated beyond 40 to 50 °C, its morphology changes, creating flow characteristics that are not just determined by temperature but also by the structure of the biomass, such as surface regression and major fragmentations that are relevant to bonding because natural binders contained in the biomass are activated upon increasing temperature [6,7,34,65]. Figure 5 presents cross-sections of the SEM images of pure and blended samples of NSP and PSP obtained under the same measurement conditions such as those inscribed in the images.

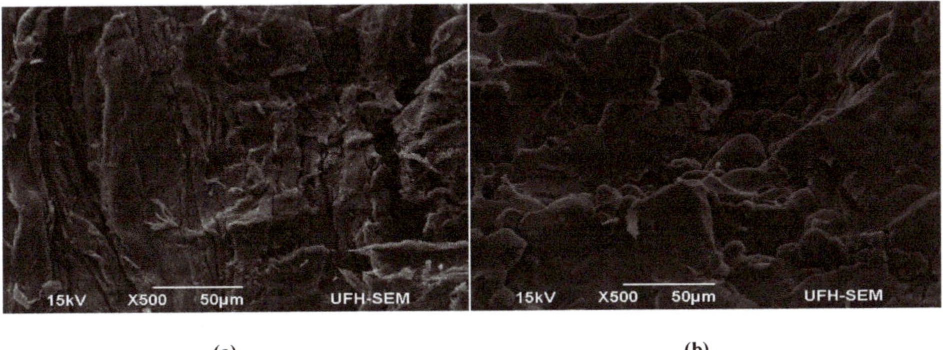

Figure 5. *Cont.*

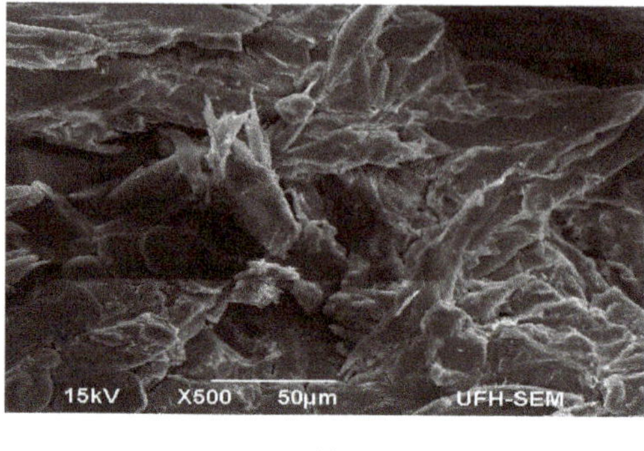

(c)

Figure 5. The SEM images of pure and blended samples of NSP and PSP: (**a**) NSP (100%); (**b**) PSP (100%); (**c**) NSP/PSP (50%/50%).

Figure 5a–c typically show the fracture surfaces of the pure and blended samples of NSP and PSP. The nature of the fracture surfaces was an indication of the production of pellets with differences in visual appearance. It could also be noted that the surfaces of the pellets were quite rough with interconnected particles and voids that characterized the image of PSP (100%), which were believed to have been created by pelleting process conditions such as compression force and temperature. The observable particle-to-particle interconnection was equally an indication of the promotion of adhesion by increased molecular contact between particles, which also represented the presence of short-range attraction forces such as intra and intermolecular forces including van der Waals forces, attributed to the presence of increased polar functional groups, free chemical bonds, and hydrogen bridges [41,66,67]. It could also be noted from the images that particles were clustered and densely packed, an indication that the materials were milled to relatively small particle sizes of <1 mm prior to pelleting. Surface morphology of the pellet samples equally showed conspicuous differences in the pattern of roughness, particularly between the images of NSP (100%) and PSP (100%), whose SEM image, as earlier stated, showed more void spaces that were believed to have been caused by a pelleting spring-back effect due to the highly reactive oxygen-containing polar functional groups associated with PSP (100%). This spring-back effect was an indication of interlocking of fibers and that bonding in 100% PSP was most likely due to dipole–dipole and hydrogen bonding. According to Mani et al. [68], the compaction of biomass grinds occurs in different stages, such as the formation of void spaces that facilitates inter-particle contacts. Surface roughness enhances bond strength, which is intimately linked to bond energy [69–71]. Therefore, the pattern of roughness is a strong indication of differences in the mechanism of bonding between the particles of the pure and blended pellet samples. However, the voids observed in the image of PSP (100%) were no longer visible in the SEM image of NSP/PSP (50%/50%), a clear evidence of component consolidation and an indication that components of the two materials (Norway spruce and pea starch) interacted to create stronger unions, perceived as a type of intermolecular bonding connected to covalent and hydrogen bonding [72]. Furthermore, the particles of PSP (100%) were more spherical in shape than those of NSP (100%), and also appeared complemented in the blend (NSP/PSP 50%/50%) (Figure 5c). Spherical particles have higher surface areas that are more densely packed; as a result, they are likely to form pellets with greater strength [49].

With regard to the quality of the pellets, the SEM images did not provide clear morphological features that could be linked to differences in the quality of the pellets, whether in terms of strength, bulk density, or of any other quality parameter for that matter; hence, it was difficult to draw conclusions on the quality of the pellets from this analysis.

4. Discussion

In this study, pure and blended fuel pellets made from Norway spruce and pea starch were analyzed using advanced analytical instruments able to provide information beyond what is visible to the naked eye in order to establish differences in particle bonding mechanism relevant to quality. The results obtained were comprehensively interpreted from a structural chemistry perspective and used to answer questions relating to how particles of pure and blended pellets made from two different biomass materials combine to form pellets, and the source of inter-particle bonding. As previously stated, assessing biomass for any conversion or pretreatment process requires an understanding of its basic composition [1,3,32]. Therefore, compositional analysis data determined that pure and blended samples of NSP and PSP assessed in this study contained various proportions of major organic and elemental constituents, which were vital to understanding the nature of the materials that were pelletized, and which were equally significant in comprehending how the major components combine to form pellets.

From the FT-IR data, it was established that the pure and blended samples of NSP and PSP contained varying concentrations of polar and non-polar functional groups that played a significant role in providing information related to differences in the mechanism of bonding and the type of attraction forces between individual particles relevant to quality in terms of strength only. It was determined that bonding in NSP (100%) was mostly due to van der Waals attraction forces, while, for PSP (100%), bonding was attributed to a combination of dipole–dipole and hydrogen bonding. In NSP/PSP (50%/50%), particles were held together by a combination of forces that included hydrogen bonding, dipole–dipole interaction, and slight van der Waals attraction forces. From previous studies [7,41,73,74], it was determined that the quality of biomass pellets, among other factors, is dependent upon the type and strength of attraction forces between individual particles; hence, FT-IR analysis aided the determination of the quality of the pellets in terms of strength based on the theory of functional groups and the strength of the forces acting between individual particles, for which the order of strength of the pellets was given in the last paragraph of Section 3.2. In this study, therefore, spectral deconvolution from FT-IR analysis provided significant information not just about the type of attraction forces between individual particles of the pellets, but also bond strength, which helped to establish the quality of the pellets without having to rely on special quality assessment tools.

Thermal analysis data from TGA and DSC of the pure and blended samples of NSP and PSP suggested that the behavior of the materials (Norway spruce and pea starch) under pelleting temperature was controlled by major components of the pellets, and that temperature differences between the pellets played a key role in determining when molecular species relevant to bonding were released. This means that how particles of the materials combined to form pellets during pelleting was determined by temperature, as noted by the degradation patterns of the resulting thermograms from TGA and DSC analyses. In relation to the quality of the pellets, which was defined in terms of burning rate, heat flow rate, and combustion efficiency, NSP (100%) dominated because of reasons given earlier in Sections 3.3 and 3.4. Differences in the modification temperatures between the pure and blended pellet samples also suggested that pelleting pure pea starch would require more energy than pelleting pure Norway spruce, and a 50/50 blend of these two materials would most likely show a balance in energy consumption during pelleting. Nevertheless, a comparative energy consumption analysis of pelleting pure and blended Norway spruce and pea starch must be conducted to establish the energy needed. Judging by the strength of the forces holding its particles together, PSP (100%) was equally a good-quality pellet in terms of strength; however, its poor thermal conductivity as a result of its high oxygen to carbon (O–C) ratio with the ability to retain moisture renders the pellet unsuitable as feedstock in thermal conversion systems for energy production purposes. Biomass feedstock with a high O–C ratio reduces the heating value of the biomass and causes an increased amount of smoke with greater formation of water vapor, as well as significant energy losses when such biomass is used as feedstock in thermal energy production systems [35,52,75–78].

From SEM analysis, it was established that the pattern of surface roughness of the pellets and the mode of interconnectivity of particles, observed from the SEM images of the pellets, offered a strong indication of differences in bonding mechanisms between the pure and blended pellet samples, which basically showed that major components from two or more different materials behave differently under certain pelleting conditions such as temperature and compression force, and that their behavior is affected by the content and concentrations of polar functional groups. According to a previous study [7], functional groups confer specific properties and act differently under certain pelleting conditions such as those presented in Table 1. With respect to the quality of the pellets, however, morphological features from SEM analysis could not establish any observable structural changes of importance to quality between the pellets, no matter how quality was defined.

Much still remains to be understood about differences in the mechanism of bonding between pure and blended pellets made from different biomass materials and the chemistry of not just the strength of the pellets, but also of their combustion. Although there are many similarities in the composition of biomass, there are equally huge differences; hence, understanding how components of different biomass combine to form good-quality pellets is made difficult by the variability in the composition of biomass and the multiple events that occur as biomass undergoes pelleting. Given this difficulty, however, it is vital to generate a full understanding of the structural and chemical properties of pelletized biomass relevant to good quality in all ramifications. As such, further studies are required on the bonding characteristics of pure and blended biomass before pelleting, employing contemporary analytical techniques so that a comparison can be made with this study in order to achieve better understanding of how milled biomass is transformed to pellet, as well as the source of inter-particle bonding in pure and blended biomass pellets made from different biomass resources. The ratio of the blended materials should also be widely varied and pelletized in order to conclusively establish differences in the mechanism of bonding relevant to quality between the varied biomass pellets. To corroborate evidence of quality of the pure and blended pellets provided by the state-of-the-art analytical instruments used in this study, it is recommended that standard quality assessment tools be used.

In light of what was investigated in this study, it is fair to allude that the significance of the study is related to the information it adds to differences in the mechanism of bonding between pure and blended biomass pellets from two different materials and how this correlates to the quality of pellets.

5. Conclusions

Based on the findings of the study, the following conclusions were drawn:

- Satisfactory information was obtained from compositional analysis of pure and blended NSP and PSP, which established that the pellets contained varying proportions of organic and elemental components that were particularly responsible for their pelleting behavior under the conditions presented in Table 1. Data from subsequent analytical instruments also established that the major components of the pellets exhibited varying bonding attributes that were used to predict their quality in terms of strength and combustion efficiency.

- The data from FT-IR analysis showed that differences in the concentration of polar functional groups between the pure and blended pellet samples explained the type and strength of attraction forces between individual particles of the pellets, with PSP (100%) exhibiting better quality in terms of strength than NSP (100%) and NSP/PSP (50%/50%) due to the strength of the forces acting between its particles. Therefore, the presence of functional groups and the variation in intermolecular forces between particles of the pure and blended pellets were the sources of differences in their particle bonding mechanism.

- Temperature was a determinant factor of how particles of the pure and blended pellet samples combined to form pellets (differences in bonding mechanism between particles of the pure and blended pellets). This was because of differences in the modification/transition temperature where molecular species relevant to bonding were considered to be released. For instance, the release of the bonding components of PSP (100%) began at a much higher temperature compared to NSP (100%) and the NSP/PSP (50%/50%) blend, a condition which suggested that flow characteristics of polymeric constituents relevant to bonding were delayed for PSP (100%), thus slowing its mechanism of particle bonding.

- Thermal analysis results from TGA and DSC also established that NSP (100%) had better burning efficiency in comparison to PSP (100%) and the NSP/PSP (50%/50%) blend.
- The variation in surface roughness and interconnectivity of particles that was noted from the SEM images of the pellets was a strong indication of differences in their particle bonding mechanism, and evidence of the application of compression force. These conditions were facilitated by the presence of functional groups, which were assumed to act differently under certain pelleting process conditions such as those presented in Table 1.
- Due to the formation of void spaces between its particles and the lack of inter-particle polymer bridges, bonding in PSP (100%) may be due to dipole–dipole forces and interlocking of fibers in comparison to van der Waals forces in NSP (100%) and hydrogen bonding in the NSP/PSP (50%/50%) blend.
- The three pellets studied all showed evidence of good quality in terms of strength; however, in terms of combustion efficiency, only two of the pellets (NSP 100% and the NSP/PSP 50%/50% blend) were considered good-quality pellets for reasons given earlier.

Author Contributions: Conceptualization, A.I.A., J.B., and S.F.; methodology, A.I.A., J.B., and S.F.; investigation, A.I.A.; data curation, A.I.A.; writing—original draft preparation, A.I.A.; writing and editing, A.I.A., J.B., and S.F.; resources, E.B.F. and P.N.

Funding: This research was funded by the Swedish Agency for Growth through the FOSBE project with grant number 20201239, and the APC was funded by Karlstad University Library.

Acknowledgments: The authors wish to thank the Division of Environmental and Energy Systems within the Department of Engineering and Chemical Sciences of Karlstad University for their administrative support.

Conflicts of Interest: The authors declare no conflicts of interest.

Abbreviation

Abbreviation	Definition
NSP	Norway spruce pellet
PSP	Pea starch pellet
TGA	Thermogravimetric analysis
DSC	Differential scanning calorimetry
EDX	Energy-dispersive X-ray spectroscopy
SEM	Scanning electron microscopy
(FT-IR)	Fourier-transform infrared spectroscopy
IR	Infrared
T_g	Glass transition

References

1. Anukam, A.; Mamphweli, S.; Reddy, P.; Meyer, E.; Okoh, O. Pre-processing of sugarcane bagasse for gasification in a downdraft biomass gasifier system: A comprehensive review. *Renew. Sustain. Energy Rev.* **2016**, *66*, 775–801. [CrossRef]
2. Ståhl, M.; Frodeson, S.; Berghel, J.; Olsson, S. Using secondary pea starch in full-scale wood fuel pellet production decreases the use of steam conditioning. In Proceedings of the World Sustainable Energy Days Conference, Wels, Austria, 27 February–1 March 2019.
3. Frodeson, S.; Henriksson, G.; Berghel, J. Pelletizing pure biomass substances to investigate the mechanical properties and bonding mechanisms. *BioResources* **2018**, *13*, 1202–1222. [CrossRef]

4. Junginger, M.; Sikkema, R.; Faaij, A. *Analysis of the Global Pellet Market: Including Major Driving Forces and Possible Technical and Non-Technical Barriers*; Copernicus Institute; Utrecht University: Utrecht, The Netherlands, 2009.

5. Ståhl, M. Improving Wood Fuel Pellets for Household Use. Perspectives on Quality, Efficiency and Environment. Ph.D. Thesis, Karlstad University, Karlstad, Sweden, 2008.

6. Kaliyan, N.; Morey, R.V. Factors affecting strength and durability of densified biomass products. *Biomass Bioenergy* **2009**, *33*, 337–359. [CrossRef]

7. Anukam, A.I.; Berghel, J.; Famewo, E.B.; Frodeson, S. Improving the understanding of the bonding mechanism of primary components in biomass pellets through the use of advanced analytical instruments. *J. Wood Chem. Technol.* **2019**. [CrossRef]

8. Frodeson, S.; Henriksson, G.; Berghel, J. Effects of moisture content during densification of biomass pellets, focusing on polysaccharide substances. *Biomass Bioenergy* **2019**, *122*, 322–330. [CrossRef]

9. Ståhl., M.; Berghel, J.; Granström, K. Improvement of wood fuel pellet quality using sustainable sugar additives. *BioResources* **2016**, *11*, 3373–3383.

10. Hosseinpourpia, R.; Echart, A.S.; Adamopoulos, S.; Gabilondo, N.; Eceiza, A. Modification of pea starch and dextrin polymers with isocyanate functional groups. *Polymers* **2018**, *10*, 939. [CrossRef] [PubMed]

11. Madsen, B.; Gamstedt, E.K. Wood versus plant fibers: Similarities and differences in composite applications. *Adv. Mater. Sci. Eng.* **2013**. [CrossRef]

12. Felhofer, M. Raman Imaging to Reveal In-situ Molecular Changes of Wood during Heartwood Formation and Drying. Master's Thesis, University of Natural Resources and Life Sciences, Vienna, Austria, 2016.

13. Fahlén, J. The Cell Wall Ultrastructure of Wood Fibres—Effects of the Chemical Pulp Fibre Line. Ph.D. Thesis, KTH Stockholm, Stockholm, Sweden, 2005.

14. Malik, B.; Pirzadah, T.B.; Islam, S.T.; Tahir, I.; Kumar, M.; ul Rehman, R. Biomass pellet technology: A green approach for sustainable development. In *Agricultural Biomass Based Potential Materials*; Hakeem, K.R., Khalid, R., Jawaid, M., Alothman, O., Eds.; Springer International Publishing: Cham, Switzerland, 2015.

15. Van Dam, J.E.G.; van den Oever, M.J.A.; Teunissen, W.; Keijsers, E.R.P.; Peralta, A.G. Process for production of high density/high performance binderless boards from wholecoconut husk. Part 1: Lignin as intrinsic thermosetting binder resin. *Ind. Crops Prod.* **2004**, *19*, 207–216. [CrossRef]

16. Granada, E.; González, L.L.M.; Míguez, J.L.; Moran, J. Fuel Lignocellulosic briquettes, die design, and products study. *Renew. Energy* **2002**, *27*, 561–573. [CrossRef]

17. Nelson, D.L.; Cox, M.M. *Lehninger Principles of Biochemistry*; Freeman, W.H., Ed.; Macmillan: New York, NY, USA, 2005.

18. Chaplin, M. Water Structure and Science—Starch. Available online: http://www1.lsbu.ac.uk/water/starch.html#r260 (accessed on 11 April 2019).

19. Buléon, A.; Colonna, P.; Planchot, V.; Ball, S. Starch granules: Structure and biosynthesis. *Int. J. Biol. Macromol.* **1998**, *23*, 85–112. [CrossRef]

20. Bertoft, E. Understanding starch structure: Recent progress. *Agronomy* **2017**, *7*, 56. [CrossRef]

21. Ratnayake, W.S.; Hoover, R.; Warkentin, T. Pea Starch: Composition, structure and properties—A Review. *Starch-Stärke* **2002**, *54*, 217–234. [CrossRef]

22. Anukam, A.I.; Berghel, J.; Henrikson, G.; Frodeson, S.; Ståhl, M. A critical review of the bonding mechanism of primary biomass components and the use of additives in densified pellets: Challenges and associated benefits. (unpublished; manuscript in preparation).

23. Henriksson, L.; Frodeson, S.; Berghel, J.; Andersson, S.; Ohlson, M. Bioresources for sustainable pellet production in Zambia: Twelve biomasses pelletized at different moisture contents. *BioResources* **2019**, *14*, 2550–2575.

24. Tumuluru, J.S. Pelleting of pine and switchgrass blends: Effect of process variables and blend ratio on the pellet quality and energy consumption. *Energies* **2019**, *12*, 1198. [CrossRef]

25. Agar, D.A.; Rudolfsson, M.; Kalén, G.; Campargue, M.; Da Silva Perez, D.; Larsson, S.H. A systematic study of ring-die pellet production from forest and agricultural biomass. *Fuel Process. Technol.* **2018**, *180*, 47–55. [CrossRef]

26. Wattana, W.; Phetklung, S.; Jakaew, W.; Chumuthai, S.; Sriam, P.; Chanurai, N. Characterization of mixed biomass pellet made from oil palm and para-rubber tree residues. *Energy Procedia* **2017**, *138*, 1128–1133. [CrossRef]

27. Wang, Y.; Wu, K.; Sun, Y. Pelletizing properties of wheat straw blending with rice straw. *Energy Fuels* **2017**, *31*, 5126–5134. [CrossRef]

28. Stelte, W.; Sanadi, A.R.; Shang, L.; Holm, J.K.; Ahrenfeldt, J.; Henriksen, U.B. Recent developments in biomass pelletization—A review. *BioResources* **2012**, *7*, 4451–4490.

29. SS-EN 14774-1. *Solid Biofuels-Determination of Moisture Content-Oven Dry Method-Part 1: Total Moisture-Reference Method*; 118 80; SIS FÖrlag AB: Stockholm, Sweden, 2009.

30. Nielsen, N.P.K.; Gardner, D.J.; Poulsen, T.; Felby, C. Importance of temperature, moisture content, and species for the conversion process of wood residues into fuel pellets. *Wood Fiber Sci.* **2009**, *41*, 414–425.

31. Nguyen, Q.N.; Cloutier, A.; Achim, A.; Stevanovic, T. Effect of process parameters and raw material characteristics on physical and mechanical properties of wood pellets made from sugar maple particles. *Biomass Bioenergy* **2015**, *80*, 338–349. [CrossRef]

32. Anukam, A.; Okoh, O.; Mamphweli, S.; Berghel, J. A comparative analysis of the gasification performances of torrefied and untorrefied bagasse: Influence of feed size, gasifier design and operating variables on gasification efficiency. *Int. J. Eng. Technol.* **2018**, *7*, 859–867. [CrossRef]

33. Anukam, A.; Mamphweli, S.; Okoh, O.; Reddy, P. Influence of torrefaction on the conversion efficiency of the gasification process of sugarcane bagasse. *Bioengineering* **2017**, *4*, 22. [CrossRef]

34. LeVan, S.L. Thermal degradation. In *Concise Encyclopedia of Wood and Wood-Based Materials*, 1st ed.; Schniewind, A.P., Arno, P., Eds.; Pergamon Press: Elmsford, NY, USA, 1989; pp. 271–273.

35. Anukam, A.I.; Mamphweli, S.N.; Reddy, P.; Okoh, O.O. Characterization and the effect of lignocellulosic biomass value addition on gasification efficiency. *Energy Exp. Exploit.* **2016**, *34*, 865–880. [CrossRef]

36. Chen, W.H.; Kuo, P.C. A study on torrefaction of various biomass materials and its impact on lignocellulosic structure simulated by a thermogravimetry. *Energy* **2010**, *35*, 2580–2586. [CrossRef]

37. Chen, W.H.; Kuo, P.C. Torrefaction and co-torrefaction characterization of hemicellulose, cellulose and lignin as well as torrefaction of some basic constituents in biomass. *Energy* **2011**, *36*, 803–811. [CrossRef]

38. Yang, H.; Yan, R.; Chen, H.; Lee, D.H.; Zheng, C. Characteristics of hemicellulose, cellulose and lignin pyrolysis. *Fuel* **2007**, *86*, 1781–1788. [CrossRef]

39. Back, E.L. The bonding mechanism in hardboard manufacture. *Holzforschung* **1987**, *41*, 247–258. [CrossRef]

40. Zheng, G.; Kozinski, J.A. Thermal events occurring during the combustion of biomass residue. *Fuel* **2000**, *79*, 181–192. [CrossRef]

41. Kaliyan, N. Binding mechanism of corn stover and switchgrass in briquettes and pellets. In Proceedings of the American Society of Agricultural and Biological Engineers Annual International Meeting, Providence, RI, USA, 29 June–2 July 2008. Abstract Number 084282.

42. Moorthy, S.N.; Andersson, L.; Eliasson, A.-C.; Santacruz, S.; Ruales, J. Determination of amylose content in different starches using modulated differential scanning calorimetry. *Starch-Stärke* **2006**, *58*, 209–214. [CrossRef]

43. Encalada, K.; Aldás, M.B.; Proaño, E.; Valle, V. An overview of starch-based biopolymers and their biodegradability. *Cienc. Ing.* **2018**, *39*, 245–258.

44. Popescu, M.C.; Popescu, C.M.; Lisa, G.; Sakata, Y. Evaluation of morphological and chemical aspects of different wood species by spectroscopy and thermal methods. *J. Mol. Struct.* **2011**, *988*, 65–72. [CrossRef]

45. LibreTexts Content. Chemistry, Infrared: Interpretation. Available online: https://chem.libretexts.org/ Bookshelves/Physical_and_Theoretical_Chemistry_Textbook_Maps/Supplemental_Modules_(Physical_ and_Theoretical_Chemistry)/Spectroscopy/Vibrational_Spectroscopy/Infrared_Spectroscopy/Infrared% 3A_Interpretation (accessed on 22 July 2019).

46. Valodkar, M.; Thakore, S. Isocyanate crosslinked reactive starch nanoparticles for thermo-responsive conducting applications. *Carbohydr. Res.* **2010**, *345*, 2354–2360. [CrossRef] [PubMed]

47. Levenson, R. *Infrared Spectroscopy: In Modern Chemical Techniques*, 3rd ed.; Pack, M.J., Berry, M., Osborne, C., Reed, N.V., Johnston, J.A., Eds.; Royal Society of Chemistry: London, UK, 1997; pp. 62–91.

48. Tumuluru, J.S.; Wright, C.T.; Hess, J.R.; Kenney, K.L. A review of biomass densification systems to develop uniform feedstock commodities for bioenergy application. *Biofuel Bioprod. Biorefin.* **2011**, *5*, 683–707. [CrossRef]

49. Tumuluru, J.S. Effect of pellet die diameter on density and durability of pellets made from high moisture woody and herbaceous biomass. *Carbon Resour. Convers.* **2018**, *1*, 44–54. [CrossRef]

50. Wolke, R.L. Bond Energy. Available online: https://science.jrank.org/pages/984/Bond-Energy.html (accessed on 22 July 2019).
51. Bahreini, M.; Movahedi, M.; Peyvandi, M.; Nematollahi, F.; Tehrani, H.S. Thermodynamics and kinetic analysis of carbon nanofibers as nanozymes. *Nanotechnol. Sci. Appl.* **2019**, *12*, 3–10. [CrossRef]
52. Anukam, A.; Mamphweli, S.; Reddy, P.; Okoh, O.; Meyer, E. An investigation into the impact of reaction temperature on various parameters during torrefaction of sugarcane bagasse relevant to gasification. *J. Chem.* **2015**. [CrossRef]
53. LibreTexts Content. Chemistry: Bond Energies. Available online: https://chem.libretexts.org/Bookshelves/Physical_and_Theoretical_Chemistry_Textbook_Maps/Supplemental_Modules_(Physical_and_Theoretical_Chemistry)/Chemical_Bonding/Fundamentals_of_Chemical_Bonding/Bond_Energies (accessed on 22 July 2019).
54. BCcampus. Chemistry. Available online: https://opentextbc.ca/chemistry/chapter/7-5-strengths-of-ionic-and-covalent-bonds/ (accessed on 22 July 2019).
55. Zafar, S. Summary of Biomass Combustion Technologies. Available online: https://www.bioenergyconsult.com/tag/biomass-combustion-process/ (accessed on 5 September 2019).
56. Gil, M.V.; Oulego, P.; Casal, M.D.; Pevida, C.; Pis, J.J.; Rubiera, F. Mechanical durability and combustion characteristics of pellets from biomass blends. *Bioresour. Technol.* **2010**, *101*, 8859–8867. [CrossRef]
57. Li, G. (Ed.) *Carbon Nanomaterials: The Fundamental Properties of Graphene and Graphene Oxide: In Nano-Inspired Biosensors for Protein Assay with Clinical Applications*, 1st ed.; Elsevier: Amsterdam, The Netherlands, 2019; Volume 1, pp. 313–330.
58. Luo, L.; Peng, T.; Yuan, M.; Sun, H.; Dai, S.; Wang, L. Preparation of graphite oxide containing different oxygen-containing functional groups and the study of ammonia gas sensitivity. *Sensors* **2018**, *18*, 3745. [CrossRef]
59. Nussbaumer, T.; Good, J. Determination of the combustion efficiency in biomass furnaces. In Proceedings of the 10th European Conference and Technology Exhibition, Würzburg, Germany, 8–11 June 1998.
60. Instructables Workshop. How to Determine Quality of Wood Pellets without Using Special Tools. Available online: https://www.instructables.com/id/5-Easy-Ways-To-Determine-Quality-Of-Wood-Pellets-W/ (accessed on 25 August 2019).
61. Huang, J. Gemco Energy: 5 Easy Ways to Determine Quality of Wood Pellets without Using Special Tools. Available online: http://www.biofuelmachines.com/determine-wood-pellet-quality-in-5-easy-ways.html (accessed on 25 August 2019).
62. Moore, J. Processing applications for roll-type briquetting-compacting machines. In Proceedings of the 9th Biennial Briquetting Conference of the International Briquetting Association, Denver, CO, USA, 2–4 September 1965.
63. Yu, L.; Christie, G. Measurement of starch thermal transitions using differential scanning calorimetry. *Carbohyd. Polym.* **2001**, *46*, 179–184. [CrossRef]
64. Mandlawy, R. Characterization of Starch Properties in Retorted Products. Master's Thesis, Chalmers University of Technology, Gothenburg, Sweden, 2013.
65. Biswas, A. Effect of Chemical and Physical Properties on Combustion of Biomass Particle. Ph.D. Thesis, Luleå University of Technology, Luleå, Sweden, 2015.
66. Chung, F.H. Unified theory and guidelines on adhesion. *J. Appl. Polym. Sci.* **1991**, *42*, 1319–1331. [CrossRef]
67. Liu, Z.; Quek, A.; Balasubramanian, R. Preparation and characterization of fuel pellets from woody biomass, agro-residues and their corresponding hydrochars. *Appl. Energy* **2014**, *113*, 1315–1322. [CrossRef]
68. Mani, S.; Tabil, L.G.; Sokhansanj, S. An overview of compaction of biomass grinds. *Powder Hand. Process.* **2003**, *15*, 160–168.
69. Budhe, S.; Ghumatkar, A.; Birajdar, N.; Banea, M.D. Effect of surface roughness using different adherend materials on the adhesive bond strength. *Appl. Adhes. Sci.* **2015**, *3*, 20. [CrossRef]
70. Uehara, K.; Sakurai, M. Bonding strength of adhesives and surface roughness of joined parts. *J. Mater. Process. Technol.* **2002**, *127*, 178–181. [CrossRef]
71. Corrosionpedia. Bond Strength. Available online: https://www.corrosionpedia.com/definition/180/bond-strength (accessed on 24 July 2019).

72. Alarcon, M.; Santos, C.; Cevallos, M.; Eyzaguirre, R.; Ponce, S. Study of the mechanical and energetic properties of pellets produce from agricultural biomass of quinoa, beans, oat, cattail and wheat. *Waste Biomass Valor.* **2017**, *8*, 2881–2888. [CrossRef]
73. Finney, K.N.; Sharifi, V.N.; Swithenbank, J. Fuel pelletization with a binder: Part I—Identification of a suitable binder for spent mushroom compost—Coal tailing pellets. *Energy Fuels* **2009**, *23*, 3195–3202. [CrossRef]
74. Stelte, W.; Holm, J.K.; Sanadi, A.R.; Barsberg, S.; Ahrenfeldt, J.; Henriksen, U.B. A study of bonding and failure mechanisms in fuel pellets from different biomass resources. *Biomass Bioenergy* **2011**, *35*, 910–918. [CrossRef]
75. Adamovics, A.; Platace, R.; Gulbe, I.; Ivanovs, S. The content of carbon and hydrogen in grass biomass and its influence on heating value. *Eng. Rural Dev.* **2018**, *23*, 1277–1281.
76. Tumuluru, J.S.; Sokhansanj, S.; Wright, C.T.; Boardman, R.D.; Hess, R.J. Review on biomass torrefaction process and product properties and design of moving bed torrefaction system model development. In Proceedings of the Annual International Meeting of the American Society of Agricultural and Biological Engineers, Louiseville, QC, Canada, 7–10 August 2011.
77. Demirbas, A. Effects of moisture and hydrogen content on the heating value of fuels. *Energy Sources Part A* **2007**, *29*, 649–655. [CrossRef]
78. Prins, M.J.; Ptasinski, K.J.; Janssen, F. More efficient biomass gasification via torrefaction. *Energy* **2006**, *31*, 3458–3470. [CrossRef]

Article

Correlations to Predict Elemental Compositions and Heating Value of Torrefied Biomass

Mahmudul Hasan [1], Yousef Haseli [1,*] and Ernur Karadogan [2]

[1] Clean Energy and Fuel Lab, School of Engineering and Technology, Central Michigan University, Mount Pleasant, MI 48859, USA; hasan2m@cmich.edu

[2] Robotics and Haptics Lab, School of Engineering and Technology, Central Michigan University, Mount Pleasant, MI 48859, USA; karad1e@cmich.edu

* Correspondence: hasel1y@cmich.edu

Received: 10 August 2018; Accepted: 10 September 2018; Published: 14 September 2018

Abstract: Measurements reported in the literature on ultimate analysis of various types of torrefied woody biomass, comprising 152 data points, have been compiled and empirical correlations are developed to predict the carbon content, hydrogen content, and heating value of a torrefied wood as a function of solid mass yield. The range of torrefaction temperature, residence time and solid yield of the collected data is 200–300 °C, 5–60 min and 58–97%, respectively. Two correlations are proposed for carbon content with a coefficient of determination (R^2) of 81.52% and 89.86%, two for hydrogen content with R^2 of 79.01% and 88.45%, and one for higher heating value with R^2 of 92.80%. The root mean square error (RMSE) values of the proposed correlations are 0.037, 0.028, 0.059, 0.043 and 0.023, respectively. The predictability of the proposed relations is examined with an additional set of experimental data and compared with the existing correlations in the literature. The new correlations can be used as a useful tool when designing torrefaction plants, furnaces, or gasifiers operating on torrefied wood.

Keywords: torrefied biomass; correlation; ultimate analysis; solid yield; heating value; OLS

---

1. Introduction

The energy demand of the world has been increasing consistently due to growing population and rising living standards [1]. As a result, fossil fuel reserves are ending gradually [2,3]. These fuels also cause severe environmental contamination with a high level of greenhouse gas emissions [4,5]. The research on renewable energy and alternative fuels has attracted many scholars worldwide. Woody biomass has been recognized as a promising source of renewable energy due to its availability in most parts of the globe [6]. Major shortcomings of biomass that limit its wider utilization for power generation are low bulk density, high moisture content, low heating value, and high energy requirement for grinding [7].

To improve the fuel properties of biomass, it needs to be pretreated. Torrefaction technology is a thermal treatment that greatly improves the heating value and grindability of biomass [8]. It is a mild pyrolysis where raw biomass loses its moisture content and a fraction of volatiles at a temperature between 200–300 °C in an inert atmosphere. The thermophysical properties of the torrefied biomass such as its elemental compositions and heating value are required for optimum design of a torrefaction plant or the furnace of a boiler which uses torrefied biomass as a fuel. Experimental determination of the elemental compositions through ultimate analysis is costly and requires sensitive equipment [9]. A predictive model that can help estimate the composition and heating value of torrefied solid obtained from different operating conditions is expected to be a useful tool for process modeling purposes. To the best knowledge of the authors, the literature lacks a general correlation for prediction of the C, H and O content of a torrefied biomass.

Some past studies have presented correlations for predicting the elemental compositions and HHV of raw biomass using proximate and ultimate analysis [10–13]. Parikh et al. [10] considered Volatile Matter (VM) and Fixed Carbon (FC) and proposed the following correlations:

$$C = 0.637FC + 0.455VM \tag{1}$$

$$H = 0.052FC + 0.062VM \tag{2}$$

$$O = 0.304FC + 0.476VM \tag{3}$$

The average absolute errors of Equations (1)–(3) are 3.21%, 4.79%, 3.4% with bias errors of 0.21%, 0.15% and 0.49%, respectively, for predicting C, H and O content of raw biomass. Shen et al. [11] developed the following relations based on fixed carbon, volatile matter and ash content (ASH) of raw biomass to predict the elemental compositions of the same raw biomass.

$$C = 0.635FC + 0.460VM - 0.095ASH \tag{4}$$

$$H = 0.059FC + 0.060VM + 0.010ASH \tag{5}$$

$$O = 0.340FC + 0.469VM - 0.023ASH \tag{6}$$

The average absolute errors of correlations given in Equations (4)–(6) are 3.17%, 4.47% and 3.16%, with average bias errors of 0.19%, 0.34% and 0.19%, respectively. Yin [12] used 44 sets of data and proposed two separate relationships for HHV of raw biomass using proximate analysis and ultimate analysis. Friedl et al. [13] proposed a correlation using 122 samples of plant materials for estimation of the HHV of raw biomass using the elemental composition. A comprehensive review of the models for estimating the heating value of biomass is reported in Ref. [14].

Due to the significant difference between the thermophysical properties of raw and torrefied biomass, the existing correlations in the literature are not suitable for estimating the elemental compositions. It is worth noting that the correlation of Friedl et al. [13] has been successfully applied for prediction of the heating value of torrefied biomass in some studies [15,16]. Nhucchen [17] used the data of proximate analysis of raw and torrefied biomass and proposed correlations for predicting the C, H and O contents of torrefied biomass. Nhucchen et al. [18] developed two correlations for the HHV of torrefied biomass using both proximate analysis and ultimate analysis. The latter correlation is alike that of Friedl et al. [13]. A model for predicting the HHV of torrefied biomass using the kinetics of decomposition with an average absolute error of 7.03% is also reported by Soponpongpipat et al. [19].

Although the proposed relations of Nhucchen et al. [17,18] appear to be useful, they do not eliminate the need for experiment. One would still need to have the measured values of the fixed carbon, volatiles, and ash. This article aims to introduce simple but accurate correlations to estimate the elemental compositions and the heating value of a torrefied biomass without a need to experimentation. The new relations developed using ordinary least squares regression model are described in terms of solid mass yield. The ultimate analyses of various types of wood, torrefied at different experimental conditions, are collected from several past studies. The proposed correlations are validated using an additional set of data. They are also compared with other correlations reported in the literature.

2. Methodology

2.1. Biomass Samples

Data related to 152 measurements were compiled from several sources that included ultimate analysis and HHV of raw and torrefied woody biomass, solid yield (Y_s), torrefaction temperature, and residence time. The data were categorized into 15 different types of wood with 43 different torrefaction conditions. Table 1 summarizes the categorized data according to the torrefaction temperature and residence time for each type of feedstock. The sources of the collected experimental data are also provided in Table 1. For instance, the torrefaction data for birch were obtained from four different sources [20–24] with a minimum and a maximum torrefaction temperature of 200 °C and 280 °C respectively, and the residence times varying between 10 and 60 min.

Table 2 presents an overview of the compiled data based on the wood type including the number of measurements (n), the minimum (Min), maximum (Max), and mean of experimentally measured Y_s, C, H, N contents, and the HHV. Table 2 does not provide the oxygen content since it can be determined by subtracting the sum of C, H and N contents from 100%. The average torrefaction temperature and residence time including all types of wood are 257 °C and 38 min, respectively. The mean of C, N, H and O contents are 53.5%, 0.17%, 5.7% and 40.5% in weight basis, respectively. The average Y_s is 78%, and the average HHV is 20.95 MJ/kg. The range of C, H, N, O contents, temperature, residence time, solid yield and HHV are 45.1–64.4%, 3.45–6.99%, 0–0.9%, 20.89–47.3%, 200–300 °C, 5–60 min, 58–97% and 16.6–26.92 MJ/kg, respectively.

Table 1. Summary of the experimental conditions of torrefaction (categorized by the feedstock type).

Feedstock Type	Temperature (°C)		Residence Time (min)		References
	Minimum	Maximum	Minimum	Maximum	
Birch	200	280	10	60	[20–24]
Pine	200	300	15	60	[22,25–30]
Spruce	200	300	5	60	[21,23,31–33]
Willow	230	300	30	60	[32,34,35]
Eucalyptus	240	290	15	60	[25,34,36]
Poplar	220	300	20	50	[27,37]
Beech Wood	240	300	15	60	[38]
Leaucaena	240	300	5	40	[39,40]
Wood Mixture	230	290	30	60	[34,41]
Lauan	220	250	30	60	[35]
Sawdust	220	300	10	60	[42]
Cedarwood	200	290	50	50	[43]
Black Locust	225	250	60	60	[44]
Ash	250	265	39	43	[32]
Aspen	240	280	15	60	[20]

2.2. Data Analysis

The observed relationships between the considered dependent and independent variables are highly linear as shown in Figures 1 and 2. Therefore, ordinary least squares regression (OLS), which is a linear regression method, was employed for analyzing the data to model the C and H content, and HHV of torrefied biomass. The assumptions made for the OLS were that the errors in the resulting prediction: (1) have an expected mean value of zero, (2) have the same variance, and (3) are not correlated with each other [45]. The correlations are developed for the HHV, carbon and hydrogen yields by excluding the nitrogen and oxygen contents as the nitrogen is assumed to remain in the solid and the oxygen content can be determined by difference. Three dimensionless parameters C_r, H_r and HHV_r defined in Equations (7)–(9), are introduced for the regression analysis.

$$C_r = \frac{C \times Y_s}{100 \times C_o} \tag{7}$$

$$H_r = \frac{H \times Y_s}{100 \times H_o} \tag{8}$$

$$HHV_r = \frac{HHV \times Y_s}{100 \times HHV_o} \tag{9}$$

where the subscript "o" refers to the raw biomass.

Table 2. Summary of the collected data (dry and ash free) categorized by the wood type.

Wood	Parameter	n	Min	Max	Mean	Wood	Parameter	n	Min	Max	Mean
Birch	Y_s (%)	19	58.01	97	76.33	Pine	Y_s (%)	24	65	95	80.53
	C (% wt.)	19	49.61	56.92	52.65		C (% wt.)	24	49.47	59.03	53.74
	N (% wt.)	19	0.06	0.15	0.12		N (% wt.)	24	0.06	0.55	0.17
	H (% wt.)	19	5.51	6.18	5.88		H (% wt.)	24	4.78	6.74	5.8
	HHV (MJ/kg)	19	18.83	22.93	20.97		HHV (MJ/kg)	24	18.07	25.38	20.96
Spruce	Y_s (%)	26	68.40	97	81.3	Eucalyptus	Y_s (%)	13	58	86.0	71.64
	C (% wt.)	26	50.6	57.49	53.89		C (% wt.)	13	50.92	63.5	55.37
	N (% wt.)	26	0.06	0.21	0.1		N (% wt.)	13	0.00	0.2	0.1
	H (% wt.)	26	5.60	6.39	5.9		H (% wt.)	13	5.3	6.31	5.85
	HHV (MJ/kg)	26	20.46	22.97	21.6		HHV (MJ/kg)	13	19.45	25	21.9
Willow	Y_s (%)	4	63	68.76	66.09	Poplar	Y_s (%)	7	73	95.0	82.14
	C (% wt.)	4	54.0	56.9	55.65		C (% wt.)	7	47.12	55.1	50.69
	N (% wt.)	4	0.00	0.42	0.16		N (% wt.)	7	0.2	0.31	0.24
	H (% wt.)	4	5.7	6.41	6.02		H (% wt.)	7	5.3	5.98	5.78
	HHV (MJ/kg)	4	21.3	23.71	22.5		HHV (MJ/kg)	7	18.5	20.8	19.6
Beech Wood	Y_s (%)	10	60	91	76.35	Leaucaena	Y_s (%)	5	60.0	79.8	67.92
	C (% wt.)	10	48.22	55.86	50.6		C (% wt.)	5	53.2	60.2	56.59
	N (% wt.)	10	0.13	0.23	0.16		N (% wt.)	5	0.7	0.9	0.81
	H (% wt.)	10	4.9	5.88	5.46		H (% wt.)	5	5.06	5.9	5.61
	HHV (MJ/kg)	10	19.01	22.00	19.93		HHV (MJ/kg)	5	21.3	24.7	22.84
Lauan	Y_s (%)	3	59.8	82.3	74.73	Sawdust	Y_s (%)	16	58.0	97	86.85
	C (% wt.)	3	54.33	64.4	57.88		C (% wt.)	16	46.9	61.0	51.16
	N (% wt.)	3	0.12	0.17	0.15		N (% wt.)	16	0.02	0.1	.06
	H (% wt.)	3	6.37	6.99	6.73		H (% wt.)	16	5.1	6.0	5.61
	HHV (MJ/kg)	3	23.2	26.92	24.45		HHV (MJ/kg)	16	16.6	26.0	18.61
Wood Mixture	Y_s (%)	14	58.8	90.5	73.07	Cedar Wood	Y_s (%)	5	69	85	77.9
	C (% wt.)	14	45.1	61.4	53.82		C (% wt.)	5	48.82	56.13	53.34
	N (% wt.)	14	0.0	0.21	0.83		N (% wt.)	5	0.48	0.83	0.61
	H (% wt.)	14	4.75	6.3	5.64		H (% wt.)	5	4.01	5.49	4.76
	HHV (MJ/kg)	14	17.8	24.3	21.34		HHV (MJ/kg)	5	19.35	21.25	20.62
Black Locust	Y_s (%)	2	79	87.0	83	Ash & Aspen	Y_s (%)	4	64.5	86.4	74.23
	C (% wt.)	2	50.59	52.77	51.68		C (% wt.)	4	50.0	53.0	51.65
	N (% wt.)	2	0.21	0.23	0.22		N (% wt.)	4	0.02	0.16	0.08
	H (% wt.)	2	3.45	4.39	3.92		H (% wt.)	4	5.9	6.1	6.0
	HHV (MJ/kg)	2	19.35	20.38	19.86		HHV (MJ/kg)	4	20.0	21.1	20.55

2.3. Error Estimation

Any empirical correlation is associated with prediction error. The accuracy of regression models in this study is examined by the root-mean-square error (RMSE), bias of the model, and the coefficient of determination (R^2) [46]. They are defined as follows:

$$\text{RMSE} = \sqrt{\frac{\sum_{i=1}^{n}\left(y_i^* - y\right)^2}{n}} \tag{10}$$

$$\text{Bias} = \frac{1}{n}\sum_{i=1}^{n}(y_i^* - y_i) \tag{11}$$

$$R^2 = 1 - \frac{\sum_{i=1}^{n}\left(y_i^* - y_i\right)^2}{\sum_{i=1}^{n}\left(y_i^* - \overline{y}\right)^2} \tag{12}$$

where y^*, y and $\overline{y}$ represent the predicted, measured and average value of the dependent variables, respectively, and n is the number of data points used for the derivation of a particular correlation. The RMSE is considered as the absolute measure of the model's fit to the data and preferred over the mean absolute error as it places more weight on the larger error terms. On the other hand, the R^2 value is the relative measure of the model's fit and represents the explained percentage of variability in the response variable as compared to the mean alone. RMSE is used for measuring the accuracy of the model along with the R^2 value when the main objective of the model is prediction. Small RMSE values and high R^2 values (between 0 and 1) imply a good fit for the model. The bias of a model measures the degree of overestimation or underestimation of the prediction as obtained from the model. Positive values of bias error suggest that the predicted values by the model will be greater than the actual values and vice versa [47].

3. Results

3.1. Elemental Compositions

Before performing a regression analysis, it is required to examine the relationship between the variables. Figure 1 shows how the C, H and N contents of torrefied woody biomass vary with solid yield using scatter plots. From Figure 1a, it is observed that the carbon content decreases linearly with the increase of solid yield for every type of feedstock. Almost every type of wood forms a data cluster. Figure 1b indicates that the hydrogen percentage increases with increasing solid yield. As the amount of nitrogen is very low, its change is not significant. However, Figure 1c shows that the nitrogen content increases with an increase in the solid yield for some of the data of pine, willow and eucalyptus, but it decreases for other types of wood. Figure 2 depicts the HHV of torrefied wood versus the carbon and hydrogen yields. A linear relation between the heating value and the carbon content is evident in Figure 2a for all types of torrefied wood. However, no distinct relation between HHV and hydrogen content can be seen in Figure 2b.

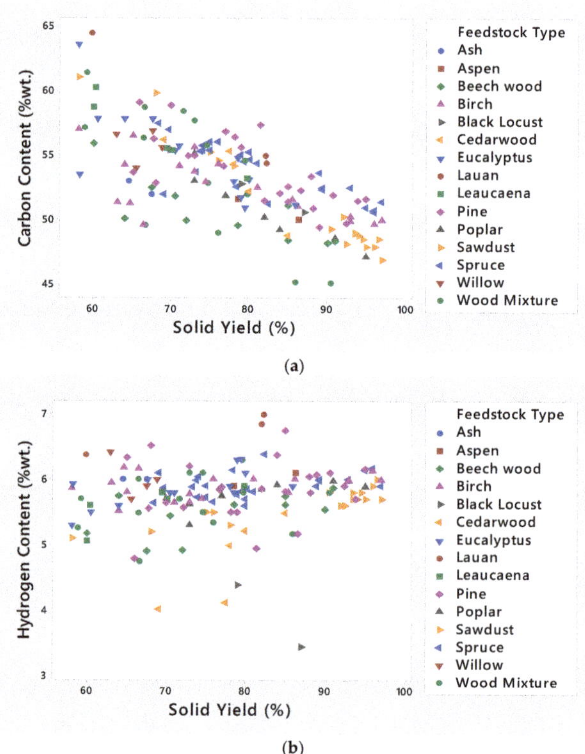

Figure 1. *Cont.*

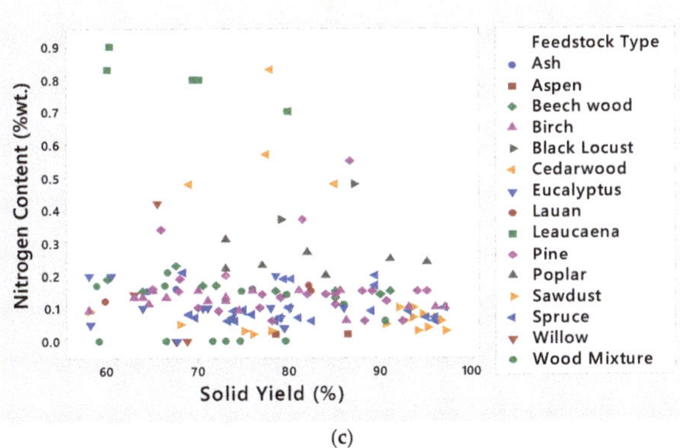

Figure 1. Variation of (**a**) Carbon (% wt.) (**b**) Hydrogen (% wt.) and (**c**) Nitrogen (% wt.) with solid yield for different types of torrefied wood.

3.2. Modeling of C, H and HHV

As mentioned previously, 152 samples have been used in regression analysis to develop correlations for C and H contents and HHV of torrefied wood. Table 3 provides five different regression equations that are evaluated for three dependent variables and their corresponding prediction measures. For instance, in the case of C_r, the first regression equation that includes only Y_s as the independent variable, Equation (13), results in an R^2 of 81.52% and RMSE of 0.037. By including C as a second independent variable, Equation (14), the R^2 value increases to 89.86% and RMSE decreases to 0.028. Similar observations are made for H_r. That is, an R^2 of 79.01% and RMSE of 0.059 are found for the model when Y_s is used as the only independent variable, Equation (15), whereas the second regression model, Equation (16), yields an R^2 value of 88.45% and RMSE of 0.043.

$$C_r = a_1 + b_1 Y_s \tag{13}$$

$$C_r = a_2 + b_2 Y_s + a_3 C \tag{14}$$

$$H_r = a_3 + b_3 Y_s \tag{15}$$

$$H_r = a_4 + b_4 Y_s + a_4 H \tag{16}$$

$$HHV_r = a_5 + b_5 C_r \tag{17}$$

Table 3. Regression models for predicting the C and H contents and HHV of torrefied wood.

Dependent Variable	Independent Variable(s)	RMSE	Bias	R^2 (%)	Eq. No.
C_r	Y_s	0.037	1.4×10^{-19}	81.52	13
	Y_s, C	0.028	1.5×10^{-19}	89.86	14
H_r	Y_s	0.059	3.2×10^{-21}	79.01	15
	Y_s, H	0.043	8.4×10^{-20}	88.45	16
HHV_r	C_r	0.023	-4.7×10^{-19}	92.80	17

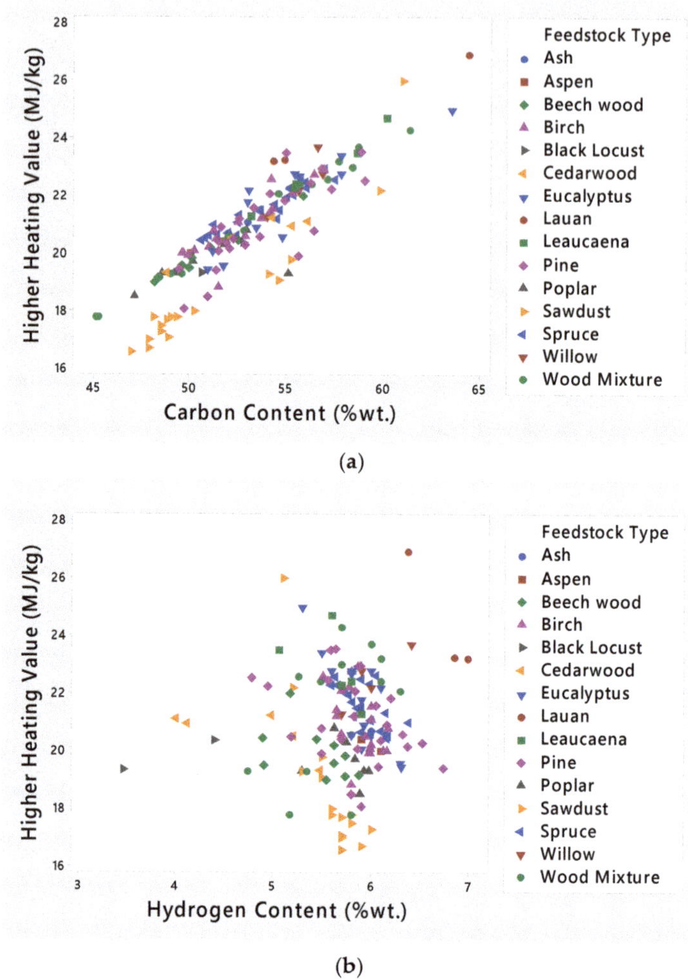

Figure 2. Relationship between HHV with (**a**) C content and (**b**) H content of torrefied wood.

For prediction of the HHV_r, an R^2 value of 92.80% with an RMSE of 0.023 is obtained by taking C_r as the only independent variable. H_r is not included as an additional variable because it does not have significance impact on HHV, which is also evident in Figure 2b. The values of the coefficients of Equations (13)–(17) in Table 3 are given in Table 4.

Table 4. The coefficients of Equations (13)–(17).

Eq. No			Eq. No			
13	a_1 0.2847	b_1 7.405×10^{-3}	14	a_2 -0.47289	b_2 9.8562×10^{-3}	c_2 1.0633×10^{-2}
15	a_3 -0.1145	b_3 1.067×10^{-2}	16	a_4 -0.55735	b_4 9.9884×10^{-3}	c_4 8.6329×10^{-2}
17	a_5 4.6508×10^{-2}	b_5 0.94497				

Figure 3 shows the deviation between the experimental and predicted values obtained from the regression models of Equations (14), (16) and (17) for C_r, H_r and HHV_r, respectively. The predicted values that are in the close vicinity of the solid lines in Figure 3a–c imply a good accuracy of prediction by the model. The data points in Figure 3a show that they form a cluster at the upper region of the solid line. The maximum points of C_r lie between 0.7 and 1. However, data for H_r are distributed through the solid line as shown in Figure 3b with 96% of the data being in the range 0.5 to 1. On the other hand, for HHV_r in Figure 3c, most of the data points are gathered in the upper region of the solid line like C_r. All the data points of HHV_r point lie between 0.65 and 1.

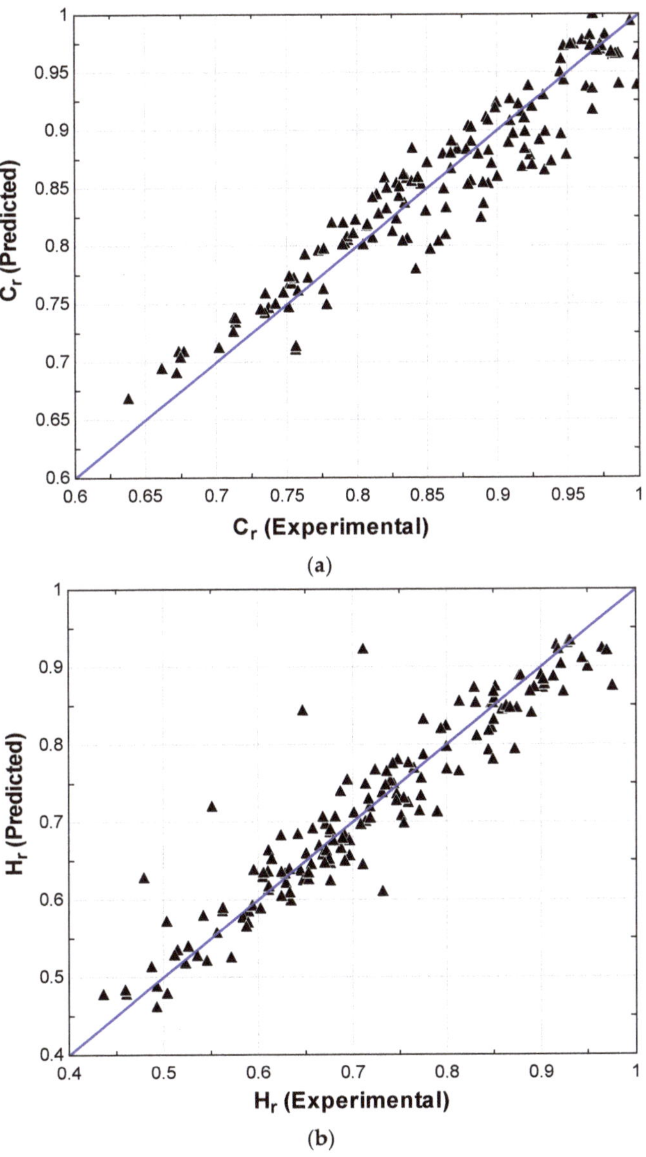

(a)

(b)

Figure 3. *Cont.*

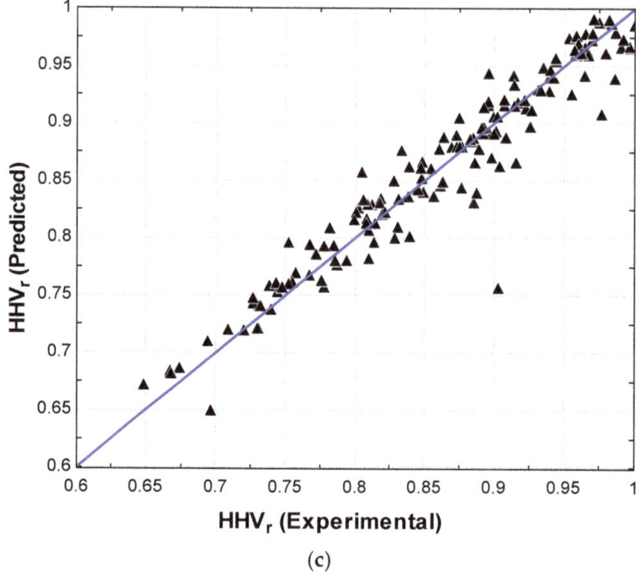

Figure 3. Regression model plot for (**a**) C_r (**b**) H_r and (**c**) HHV_r.

Introducing Equations (7)–(9) into Equations (13)–(17) and using the values of the coefficients given in Table 4, the final form of the correlations for carbon and hydrogen contents and HHV of a torrefied biomass can be represented as:

$$\frac{C}{C_o} = 0.7405 + \frac{28.47}{Y_s} \tag{18}$$

$$\frac{C}{C_o} = \frac{-0.47289 + 9.8562 \times 10^{-3} Y_s}{0.01 Y_s - 0.010633 C_o} \tag{19}$$

$$\frac{H}{H_o} = 1.067 - \frac{11.45}{Y_s} \tag{20}$$

$$\frac{H}{H_o} = \frac{-0.55735 + 9.9884 \times 10^{-3} Y_s}{0.01 Y_s - 0.086329 H_o} \tag{21}$$

$$\frac{HHV}{HHV_o} = \frac{4.6508}{Y_s} + 0.94497 \frac{C}{C_o} \tag{22}$$

where C, H and Y_s are expressed on a dry mass percentage basis.

3.3. Comparison with the Existing Correlations in the Literature and Experimental Data

As discussed previously, the proposed correlations in the literature are based on proximate analysis of raw biomass. The predictability of the correlations developed in this study are compared with those proposed by others. Parikh et al. [10] and Shen et al. [11] developed equations based on proximate analysis to predict elemental composition of raw biomass. Nhucchen [17] used the proximate analysis to predict the C and H percentage of both raw and torrefied biomass. To compare the HHV model, five different correlations from three different sources [12,13,18] are selected. Yin [12] and Friedl et al. [13] correlations are based on the proximate and ultimate analysis, whereas Nhucchen et al. [18] estimates the HHV of torrefied biomass using either proximate or ultimate analysis of torrrefied biomass.

Table 5 compares the proposed correlations for C, H and HHV in the current study to other published correlations. Average biased error (ABE) and root-mean-square error (RMSE) are calculated for all models using additional 12 samples obtained from Refs. [48,49]. Bridgeman et al. [48] torrefied willow and miscanthus at 240 °C and 290 °C with a residence time of 10 and 60 min. Broström et al. [49] reported the ultimate analysis of spruce torrefied at 260, 280 and 310 °C and a residence time of 8, 16.5 and 25 min. Although, miscanthus is a non-woody biomass, it is considered here to examine the applicability of the correlations developed in this study to non-woody biomass.

Table 5. Comparison of the newly developed C, H and HHV correlations with others.

Correlation	Reference	ABE [a] (%)	RMSE
$\frac{C}{C_o} = 0.7405 + \frac{28.47}{Y_s}$ $\frac{C}{C_o} = \frac{-0.47289 + 9.8562 \times 10^{-3} Y_s}{0.01 Y_s - 0.010633 C_o}$	Current Study	-2.4 4.0	2.12 3.30
$C = 0.637 FC + 0.455 VM$	[10]	-12.1	7.39
$C = 0.635 FC + 0.460 VM - 0.095 ASH$	[11]	-11.7	7.23
$C = -35.9972 + 0.7698 VM + 1.3269 FC + 0.3250 ASH$	[17]	-6.1	3.86
$\frac{H}{H_o} = 1.067 - \frac{11.45}{Y_s}$ $\frac{H}{H_o} = \frac{-0.55735 + 9.9884 \times 10^{-3} Y_s}{0.01 Y_s - 0.086329 H_o}$	Current Study	-2.4 -4.8	0.24 0.88
$H = 0.052 FC + 0.062 VM$	[10]	-1.8	0.26
$H = 0.059 FC + 0.060 VM + 0.010 ASH$	[11]	-1.4	0.26
$H = 55.3678 - 0.4830 VM - 0.5319 FC - 0.5600 ASH$	[17]	11.9	0.82
$\frac{HHV}{HHV_o} = \frac{4.6508}{Y_s} + 0.94497 \frac{C}{C_o}$	Current Study	-3.1 [b] 2.93 [c]	1.02 1.24
$HHV = 0.1905 VM + 0.2521 FC$	[12]	-8.7	2.32
$HHV = 0.2949 C + 0.8250 H$	[12]	-3.2	0.92
$HHV = 3.55 C^2 - 232 C - 2230 H + 51.2 CH + 131 N + 20600$	[13]	1.3	0.44
$HHV = 0.1846 VM + 0.3525 FC$	[18]	-0.6	0.78
$HHV = 32.7934 + 0.0053 C^2 - 0.5321 C - 2.8769 H + 0.0608 CH - 0.2401 N$	[18]	2.2	0.55

[a] $ABE = \frac{1}{n} \sum_{i=1}^{n} (y_i^* - y_i)/y_i$, [b] carbon content calculated using Equation (18), [c] carbon content calculated using Equation (19).

The first model developed in this study for carbon content, Equation (18), gives an ABE of -2.4% and RMSE of 2.12 whereas the second correlation, Equation (19), yields an ABE of 4% and RMSE of 3.3. The correlations of Parikh et al. [10] and Shen et al. [11] give higher negative values of -12.1% and -11.7% for ABE with RMSE of 7.39 and 7.23, respectively, as their models are originally developed for raw biomass. The correlation proposed by Nhucchen [17] shows an ABE of -6.1% and RMSE of 3.86. In the case of hydrogen, the correlations of this study, Equations (20) and (21), show an ABE of -2.4% and -4.8% with an RMSE of 0.24 and 0.88, respectively. The highest ABE is found for the correlation of Nhucchen [17]. The correlation developed by Prikh et al. [10] gives an ABE of -1.8% and RMSE of 0.26 and that of Shen et al. [11] yields an ABE of -1.4% and RMSE of 0.26.

For HHV, this study shows an ABE of -3.1% and RMSE of 1.02 when Equation (18) is used to determine the carbon content. These figures slightly change to 2.93% and 1.24 if the carbon content is calculated using Equation (19). Yin et al. [12] developed two correlations for HHV of raw biomass. As shown in Table 5, the first one gives an ABE of -8.7% and RMSE of 2.32 whereas the second relation yields an ABE of -3.2% and RMSE of 0.92. Friedl et al. correlation [13] developed using raw biomass data shows an ABE of 1.3% and RMSE of 0.44. The two correlations developed by Nhucchen et al. [18] give ABE of -0.6% and 2.2% with RMSE of 0.78 and 0.55.

Figure 4 compares the predicted values of the carbon content, the hydrogen content, and heating value obtained from Equations (18)–(22) with the measured data reported in Refs. [48,49]. It is evident from Figure 4a that Equations (18) and (19) satisfactorily predict the C content. A maximum error of 7.23% is found for miscanthus when C is predicted by Equation (18). For woody biomass, a maximum error of 8.6% is found for spruce from Equation (19). In the case of H, as shown in Figure 4b, the largest difference between the predicted and measured values found for miscanthus is 12.33% when Equation (21) is used. Also, a reasonable accuracy of prediction is found for HHV in Figure 4c with a maximum error of 8.2%. Overall, the agreement between the predicted and measured values in Figure 4 is acceptable for engineering applications.

It must be noted that as the proposed correlations are obtained using the data of different types of wood, their application to a non-woody biomass is not recommended. Furthermore, the new correlations are developed for a solid yield in the range 58% to 97%. Whether torrefied biomass is used in a combustion or gasification process, one would need to know the composition of the torrefied solid to accurately predict the gaseous products (combustion gases or producer gas). For this, Equations (18)–(22) are expected to be a useful tool for designers and researchers.

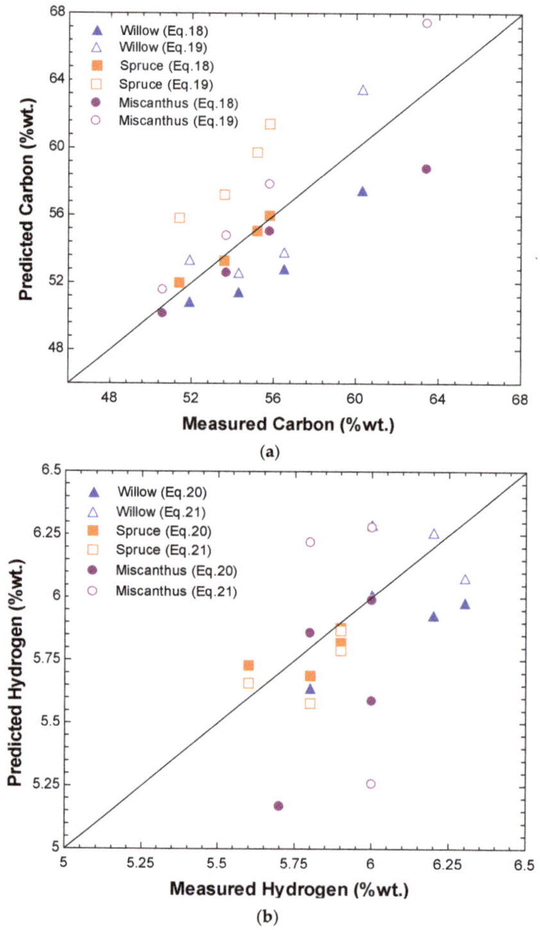

Figure 4. *Cont.*

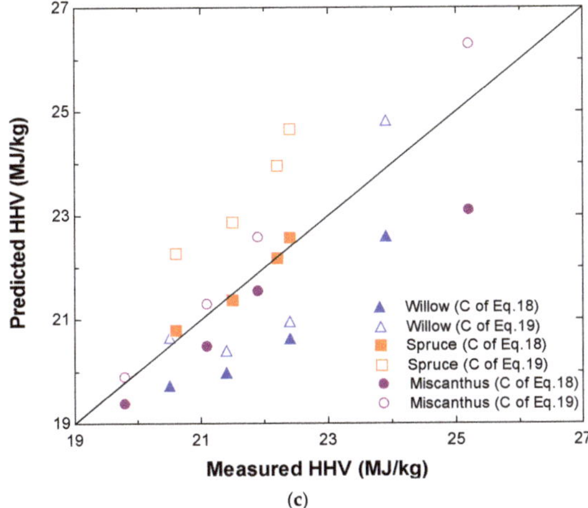

Figure 4. Comparison of the predicted and measured values of (**a**) Carbon content, (**b**) Hydrogen content and (**c**) HHV.

4. Conclusions

New correlations are developed using the ultimate analyses of torrefied woody biomass comprising 152 data points and 15 different types of wood, for predicting the carbon content, hydrogen content, and heating value of a torrefied wood. The proposed relations given in Equations (18) through (22) allow estimation of the C and H contents and HHV of a torrefied wood as a function of solid mass yield, composition and heating value of untreated wood. The accuracy of the proposed correlations is examined using additional experimental data related to torrefaction of willow, miscanthus, and spruce. Overall, the agreement between the predicted and measured quantities of the carbon, hydrogen and HHV is satisfactory. The proposed correlations provide a useful tool that can be incorporated in process models of energy systems operating on torrefied wood.

Author Contributions: Conceptualization, Y.H.; Methodology, Y.H.; Validation, M.H.; Formal Analysis, M.H. and Y.H.; Investigation, M.H. and Y.H.; Data Curation, M.H.; Writing-Original Draft Preparation, M.H.; Writing-Review & Editing, Y.H. and E.K.; Supervision, Y.H. and E.K.

Funding: This research received no external funding.

Acknowledgments: The funding support of Central Michigan University is gratefully acknowledged.

Conflicts of Interest: The authors declare no conflict of interest.

Nomenclature and Subscript

C	Carbon content (%wt.)
O	Oxygen content (%wt.)
HHV	Higher heating value (MJ/kg)
VM	Volatile matter (%)
Y_s	Solid mass yield (%)
y^*	Predicted value
$\bar{y}$	Average value
daf	Dry and ash free
RMSE	Root mean square error
H	Hydrogen content (%wt.)
N	Nitrogen content (%wt.)

FC	Fixed carbon (%)
ASH	Ash content (%)
n	Number of measurements
y	Measured value
N	Number of data
ABE	Average biased error (%)
o	Raw biomass

References

1. Chen, W.; Peng, J.; Bi, X. A state-of-the-art review of biomass torrefation, densification and applications. *Renew. Sustain. Energy Rev.* **2015**, *44*, 847–866. [CrossRef]
2. Mohr, S.H.; Wang, J.; Ellem, G.; Ward, J.; Giurco, D. Projection of world fossil fuels by country. *Fuel* **2015**, *141*, 120–135. [CrossRef]
3. Shafiee, S.; Topal, E. When will fossil fuel reserves be diminished? *Energy Policy* **2009**, *37*, 181–189. [CrossRef]
4. Madanayake, B.N.; Gan, S.; Eastwick, C.; Ng, H.K. Biomass as an energy source in coal co-firing and its feasibility enhancement via pre-treatment techniques. *Fuel Process. Technol.* **2017**, *159*, 287–305. [CrossRef]
5. Herbert, G.M.J.; Krishnan, A.U. Quantifying environmental performance of biomass energy. *Renew. Sustain. Energy Rev.* **2016**, *592*, 92–308.
6. Van der Stelt, M.J.C.; Gerhauser, H.; Kiel, J.H.A.; Ptasinski, K.J. Biomass upgrading by torrefaction for the production of biofuel: A Riview. *Biomass Bioenergy* **2011**, *35*, 3748–3762. [CrossRef]
7. Haseli, Y. Process modeling of a biomass torrefaction plant. *Energy Fuels* **2018**, *32*, 5611–5622. [CrossRef]
8. Repellin, V.; Govin, A.; Rolland, M.; Guyonnet, R. Energy requirement for fine grinding of torrefied wood. *Biomass Bioenergy* **2010**, *34*, 923–930. [CrossRef]
9. Demirbas, A. Calculation of higher heating values of biomass fuel. *Fuel* **1997**, *76*, 431–434.
10. Parikh, J.; Channiwala, S.A.; Ghosal, G.K. A correlation for calculating elemental composition from proximate analysis of biomass materials. *Fuel* **2007**, *86*, 1710–1719. [CrossRef]
11. Shen, J.; Zhu, S.; Liu, X.; Zhang, H.; Tan, J. The prediction of elemental composition of biomass based on proximate analysis. *Energy Convers. Manag.* **2010**, *51*, 983–987. [CrossRef]
12. Yin, C.Y. Prediction of higher heating values of biomass from proximate. *Fuel* **2011**, *90*, 1128–1132. [CrossRef]
13. Friedl, A.; Padouvas, E.; Rotter, H.; Varmuza, K. Prediction of heating values of biomass fuel from elemental composition. *Anal. Chim. Acta* **2005**, *544*, 191–198. [CrossRef]
14. Vargas-Morenoa, J.M.; Callejón-Ferrea, A.J.; Pérez-Alonsoa, J.; Velázquez-Martí, B. A review of the mathematical models for predicting the heating value of biomass materials. *Renew. Sustain. Energy Rev.* **2012**, *16*, 3065–3083. [CrossRef]
15. Bridgeman, T.G.; Jones, J.M.; Shield, I.; Williams, P.T. Torrefaction of reed canary grass, wheat straw and willow to enhance solid fuel qualities and combustion properties. *Fuel* **2008**, *87*, 844–856. [CrossRef]
16. Joshi, Y.; de Vries, H.; Woudstra, T.; de Jong, W. Torrefaction: Unit operation modelling and process simulation. *Appl. Therm. Eng.* **2015**, *74*, 83–88. [CrossRef]
17. Nhucchen, D.R. Prediction of carbon, hydrogen, and oxygen compositions of raw and torrefied biomass using proximate analysis. *Fuel* **2016**, *180*, 348–356. [CrossRef]
18. Nhuccen, D.R.; Afzal, M.T. HHV Predicting Correlations for Torrefied Biomass Using Proximate and Ultimate Analyses. *Bioengineering* **2017**, *4*, 7. [CrossRef] [PubMed]
19. Soponpongpipat, N.; Sittikul, D.; Comsawang, P. Prediction model of higher heating value of torrefied biomass based on the kinetics of biomass decomposition. *J. Energy Inst.* **2015**, *89*, 425–435. [CrossRef]
20. Sarvaramini, A.; Assima, G.P.; Larachi, F. Dry torrefaction of biomass-Torrefied products and torrefaction kinetics using the distributed activation energy model. *Chem. Eng. J.* **2013**, *229*, 498–507. [CrossRef]
21. Bach, Q.V.; Tran, K.Q.; Skreiberg, O.; Trinh, T.T. Effects of wet torrefaction on pyrolysis of woody biomass fuels. *Energy* **2015**, *88*, 443–456. [CrossRef]
22. Pach, M.; Zanzi, R.; Björnbom, E. Torrefied Biomass a Substitute for Wood and Charcoal. In Proceedings of the 6th Asia-Pacific International Symposium on Combustion and Energy Utilization, Kuala Lumpur, Malaysia, 20–22 May 2002.

23. Tapasvi, D.; Khalil, R.; Tran, K.Q. Torrefaction of Norwegian Birch and Spruce: An Experimental Study Using Macro-TGA. *Energy Fuels* **2012**, *26*, 5232–5240. [CrossRef]
24. Shoulaifar, T.K.; DeMartini, N. Ash-Forming Matter in Torrefied Birch Wood: Changes in Chemical Association. *Energy Fuels* **2013**, *27*, 5684–5690. [CrossRef]
25. Arteaga-Pérez, L.E.; Segura, C. Torrefaction of Pinus radiata and Eucalyptus globulus: A combined experimental and modeling approach to process synthesis. *Energy Sustain. Dev.* **2015**, *29*, 13–23. [CrossRef]
26. Phanphanich, M.; Mani, S. Impact of torrefaction on the grindability and fuel characteristics of forest biomass. *Bioresour. Technol.* **2011**, *102*, 1246–1253. [CrossRef] [PubMed]
27. Li, M.F.; Chen, L.X. Evaluation of the structure and fuel properties of lignocelluloses through carbon dioxide torrefaction. *Energy Convers. Manag.* **2016**, *119*, 463–472. [CrossRef]
28. Carmona, S.R.; Oerez, J.F. Effect of torrefaction temperature on properties of Patula pine. *Maderas-Cienc. Tecnol.* **2017**, *19*, 39–50.
29. McNamee, P.; Adams, P.W.R.; McManus, M.C. An assessment of the torrefaction of North American pine and life cycle greenhouse gas emissions. *Energy Convers. Manag.* **2016**, *113*, 17–188. [CrossRef]
30. Zheng, Y.; Tao, L.; Yang, X. Effect of the torrefaction Temperature on structural properties and pyrolysis behavior of biomass. *BioResource* **2017**, *12*, 3425–3447. [CrossRef]
31. Grigiante, M.; Antolini, D. Mass yield as guide parameter of the torrefaction process. An experimental study of the solid fuel properties referred to two types of biomass. *Fuel* **2015**, *153*, 499–509. [CrossRef]
32. Nanou, P.; Carbo, M.C.; Keil, J. Detailed mapping of the mass and energy balance of a continuous biomass torrefaction plant. *Biomass Bioenergy* **2016**, *89*, 67–77. [CrossRef]
33. Larsson, S.H.; Rudolfsson, M. Effects of moisture content, torrefaction temperature, and die temperature in pilot scale pelletizing of torrefied Norway spruce. *Appl. Energy* **2013**, *102*, 827–832. [CrossRef]
34. Ibrahim, R.H.H.; Dravell, L.I.; Jones, J.M.; Williams, A. Physicochemical characterisation of torrefied biomass. *J. Anal. Appl. Pyrolysis* **2013**, *103*, 21–30. [CrossRef]
35. Chen, W.H.; Cheng, W.Y.; Lu, K.M.; Huang, Y.P. An evaluation on improvement of pulverized biomass property for solid fuel through torrefaction. *Appl. Energy* **2011**, *88*, 3636–3644. [CrossRef]
36. Arias, B.; Pevida, C.; Fermoso, J.; Plaza, M.G.; Rubiera, F.; Pis, J.J. Influence of torrefaction on the griandability and reactivity of woody biomass. *Fuel Process. Technol.* **2008**, *89*, 169–175. [CrossRef]
37. Na, B.; Ahn, B.J.; Lee, J.W. Changes in chemical and physical properties of yellow poplar (Liriodendron tulipifera) during torrefaction. *Wood Sci. Technol.* **2015**, *49*, 257–272. [CrossRef]
38. Gucho, E.M.; Shahzad, K.; Bramer, E.A.; Brem, G. Experimental Study on Dry Torrefaction of Beech Wood and Miscanthus. *Energies* **2015**, *8*, 3903–3923. [CrossRef]
39. Wannapeera, J.; Worasuwannarak, N. Examinations of chemical properties and pyrolysis behaviors oftorrefied woody biomass prepared at the same torrefactionmass yields. *J. Anal. Appl. Pyrolysis* **2015**, *115*, 279–287. [CrossRef]
40. Huang, Y.F.; Sung, H.T.; Chiueh, P.T.; Lo, S.L. Microwave torrefaction of sewage sludge and leucaena. *J. Taiwan Inst. Chem. Eng.* **2017**, *70*, 236–243. [CrossRef]
41. Wilk, M.; Magdziarz, A.; Kalemba, I. Characterisation of renewable fuels' torrefaction process with different instrumental techniques. *Energy* **2015**, *87*, 259–269. [CrossRef]
42. Yoo, H.S.; Choi, H.S. A study on torrefaction characteristics of waste sawdust in an auger type pyrolyzer. *J. Mater. Cycles Waste Manag.* **2016**, *18*, 460–468. [CrossRef]
43. Mei, Y.; Liu, R.; Yang, Q.; Yang, H.; Shao, J.; Draper, C.; Zhang, S.; Chen, H. Torrefaction of cedarwood in a pilot scale rotary kiln and the influence of industrial flue gas. *Bioresour. Technol.* **2015**, *177*, 355–360. [CrossRef] [PubMed]
44. Barta-Rajnai, E.; Jakab, E.; Sebestyen, Z.; May, Z.; Barta, Z.; Wang, L.; Skreiberg, O.; Gronil, M.; Bozi, J.; Czegeny, Z. Comprehensive Compositional Study of Torrefied Wood and Herbaceous Materials by Chemical Analysis and Thermoanalytical Methods. *Energy Fuels* **2016**, *30*, 8019–8030. [CrossRef]
45. Montgomery, D.C.; Peck, E.A.; Vining, G.G. *Introduction to Linear Regression Analysis*, 5th ed.; John Wiley & Sons, Incorporated: New York, NY, USA, 2012.
46. Comrie, A.C. Comparing Neural Networks and Regression Models for Ozone Forecasting. *J. Air Waste Manag. Assoc.* **1997**, *47*, 653–663. [CrossRef]
47. Willmott, C.J. Some Comments on the Evaluation of Model Performance. *Am. Metrol. Soc.* **1982**, *63*, 1309–1317. [CrossRef]

48. Bridgeman, T.G.; Jones, J.M.; Williams, A.; Waldron, D.J. An investigation of the grindability of two torrefied energy crops. *Fuel* **2010**, *89*, 3911–3918. [CrossRef]
49. Brostrom, M.; Nordin, A.; Pommer, L.; Branca, C.; Blasi, C.D. Influence of torrefaction on the devolatilization and oxidation kinetics of wood. *J. Anal. Appl. Pyrolysis* **2012**, *96*, 100–109. [CrossRef]

 *energies*

Article

Impact of Thermal Pretreatment Temperatures on Woody Biomass Chemical Composition, Physical Properties and Microstructure

Ping Wang * and Bret H. Howard

Department of Energy (DOE), National Energy Technology Laboratory (NETL), 626 Cochrans Mill Road, Pittsburgh, PA 15236, USA; bret.howard@netl.doe.gov
* Correspondence: ping.wang@netl.doe.gov; Tel.: +1-412-386-7539

Received: 29 September 2017; Accepted: 19 December 2017; Published: 23 December 2017

Abstract: Thermal pretreatment of biomass by torrefaction and low temperature pyrolysis has the potential for generating high quality and more suitable fuels. To utilize a model to describe the complex and dynamic changes taking place during these two treatments for process design, optimization and scale-up, detailed data is needed on the property evolution during treatment of well-defined individual biomass particles. The objectives of this study are to investigate the influence of thermal pretreatment temperatures on wood biomass biochemical compositions, physical properties and microstructure. Wild cherry wood was selected as a model biomass and prepared for this study. The well-defined wood particle samples were consecutively heated at 220, 260, 300, 350, 450 and 550 °C for 0.5 h under nitrogen. Untreated and treated samples were characterized for biochemical composition changes (cellulose, hemicellulose, and lignin) by thermogravimetric analyzer (TGA), physical properties (color, dimensions, weight, density and grindablity), chemical property (proximate analysis and heating value) and microstructural changes by scanning electron microscopy (SEM). Hemicellulose was mostly decomposed in the samples treated at 260 and 300 °C and resulted in the cell walls weakening resulting in improved grindability. The dimensions of the wood were reduced in all directions and shrinkage increased with increased treatment temperature and weight loss. With increased treatment temperature, losses of weight and volume increased and bulk density decreased. The low temperature pyrolyzed wood samples improved solid fuel property with high fuel ratio, which are close to lignite/bituminous coal. Morphology of the wood remained intact through the treatment range but the cell walls were thinner. These results will improve the understanding of the property changes of the biomass during pretreatment and will help to develop models for process simulation and potential application of the treated biomass.

Keywords: pyrolysis; chemical composition; micro-structure; physical properties; scanning electron microscopy; wood; thermal pretreatment; torrefaction

---

1. Introduction

Biomass is a renewable fuel and carbon natural because it consumes CO_2 from the atmosphere during growth. CO_2 as a primary greenhouse gas (GHG) is widely believed to be a major contributor to climate change. Reducing CO_2 emissions is the main advantage of utilizing biomass. Biomass can be converted to gas or liquid fuels (using gasification, anaerobic digestion, pyrolysis, fermentation and transesterification), heat and power, and chemicals. Heat and power generation for industrial and utility-scale applications primarily use direct combustion or co-firing (replacing a portion of the coal with biomass in coal-fired plants) [1]. Recently, the United Kingdom started using wood pellets in co-firing and in dedicated biomass power plants to meet the European Commission's 2020 climate and energy plan, which primarily is a reduction in GHG and an increase of renewable energy usage in total energy consumption [2]. In the U.S, biomass (wood and waste) was used to generate about 1.5% of the total power output in 2016 [3]. To increase domestic biomass utilization, the development of technologies to reduce its cost and increase its utilization efficiency is required.

Developing large-scale (>50 MW) biomass power plants and co-firing may allow for power generation at high efficiencies and relatively low costs [1]. In addition, reducing biomass feedstock processing and transportation costs might help to further lower power generation cost. Biomass is generally considered a low quality fuel mainly due to its lower energy density which is attributed to a high moisture content, less carbon, more oxygen, lower density and lower heating value [4]. These characteristics contribute to inefficiencies associated with transportation, handling, storage and conversion of biomass in an efficient and economic manner. High transportation cost results in less biomass being collected and utilized. Biomass combustion yields a lower flame temperature which results in decreased thermal efficiency [5]. Gasification of biomass results in a lower quality syngas with a high tar concentration [6]. For a pulverized coal (PC) boiler, used by the majority of coal power plants, or an entrained-flow gasifier (used in integrated gasification combined cycle (IGCC)), the average coal particle size is required to be less than 100 μm (0.1 mm) [7,8]. Reducing biomass particle size to below 0.2 mm without pretreatment is difficult and costly because biomass is fibrous and compressible [9]. Torrefaction and low temperature pyrolysis carbonization processes have been proposed as pretreatment methods that could potentially address these issues [5,10–15].

Torrefaction has been extensively investigated in recent years and was reviewed by Madanayake [16], Chen [15], Chew [14] and Van der Stelt [13]. It has been successfully tested and demonstrated in pilot scale and (semi)commercial facilities [17]. Torrefaction is a mild pyrolysis which takes place at 200 to 300 °C in an inert or non-oxidative environment with the aim of producing torrefied biomass (solid) as the primary product [13,15,18,19]. It focuses on improving physicochemical properties of biomass including increased energy density, improved grindability, and higher hydrophobicity [5,18,20,21]. The heating value of torrefied woody biomass can be increased by 37% compared to the untreated wood [16,22]. For wood treated at 240 °C for 0.5 h and ground, the percentage of particles less than 415 μm (0.415 mm) and 150 μm (0.150 mm) were double that of untreated wood [22]. Torrefaction research has been focused on experimental studies of process conditions (temperature, gas environment and time), comparison of various biomass species and their major components, characterization of the products, and applying the torrefied biomass in densification, co-firing in coal power plant, gasification and ironmaking [15,16]. Using torrefied biomass pellets could improve gasification in terms of both energy efficiency and syngas quality because of the removal of oxygenated volatile compounds [15,23]. In addition to non-oxidative or conventional torrefaction, researchers have recently investigated oxidative torrefaction, wet torrefaction, and steam torrefaction to develop alternative technologies to upgrade biomass [15].

Low temperature pyrolysis with the aim of producing biochar (solid) as the primary product occurs in the range of 300 to 500 °C [10,12,19]. It significantly improves fuel combustion properties with increased thermal conversion efficiency [5,24]. It generates biochar with higher carbon content (FC-fixed carbon) and fuel ratio (FC/VM-volatile matter) compared to torrefied biomass [19]. Chars produced by torrefaction at 300 °C or pyrolysis below 500 °C have fuel properties such as fuel ratio, burnout and ignition temperature that fall between a high-volatile and a low-volatile bituminous coal [19]. Biochar produced at 350 °C had higher combustion rate than the torrefied woody biomass at 275 °C [12]. Treatment temperature is the key difference defining torrefaction and low temperature pyrolysis. The temperature ranges of the reported studies were 300–500 °C [10], 200–400 °C [12], 200–330 °C [5], and 250–300/400–500 °C [19]. Fundamental studies spanning the temperature ranges of both of these processes (200 to 500 °C) are lacking, therefore, the primary focus of this study is to address this deficiency.

Computer modeling is the primary approach along with experimentation to help with fuel conversion process design, optimization and scale-up [25–27]. Experimental studies are necessary to understand the processes and to obtain experimental data needed to develop and validate the models. To completely describe the complex and dynamic process by a model, a biomass particle with controlled, well documented characteristics is used to follow the evolution of the individual particles through pyrolysis [25], combustion [26] and gasification [27]. A similar experimental method was applied for this pretreatment study. The most commonly reported experimental studies have investigated small amounts of milled small particles (for example <500 μm (<0.500 mm), 15 g) [22,24,28,29] or larger amounts of chips/blocks (for example 10 mm × 20 mm × 3 mm, 80 g) [10,12,20].To address the lack of fundamental data, this study utilized well defined wood particle samples to obtain torrefaction and low temperature pyrolysis data to thoroughly understand the processes and aid future modeling studies. Physical (appearance, weight, size, grindability) and chemical properties (chemical composition and proximate analysis), and heating value of the untreated and treated woods were investigated. The variations in shrinkage in the three sample reference directions are included because the shrinkage affects heat transfer to the particle and gas flow within the particle [25]. The shrinkage as a function of the mass loss was addressed in this study because the shrinkage is caused by loss of water and a portion of the mass of the biomass due to decomposition to volatiles. These parameters are important for modeling studies.

Wood and wood residues are one of the major sources of biomass [30]. Wood is essentially a series of elongated tubular fibers or cells aligned with the axis or longitudinal direction of the tree trunk and cemented together [31,32]. Each cell wall is composed of various quantities of three polymers: cellulose, hemicellulose, and lignin [32]. Cellulose is the basic structural component of all wood cell walls and primarily responsible for the strength of wood. Hemicellulose acts as a matrix for the cellulose and increases the packing density of the cell walls. Lignin acts as a glue which holds wood fibers together. Thermal treatments of wood cause the thermal degradation of the three cell wall polymers and result in physical and chemical changes of the wood. The physical changes in size, density, and grindability impact supply, handling and conversion of wood. The changes in biochemical composition significantly affect the wood properties and its conversion process. The high heating values of lignin are reported to be higher than the cellulose and hemicellulose due to higher degree of oxidation of the cellulose and hemicellulose [4]. The air-steam gasification conversions of cellulose, hemicellulose and lignin on a carbon basis are 97.9%, 92.2%, and 52.8%, respectively [33]. The product gas composition from cellulose in mol % is 35.5% CO, 27.0% CO_2, and 28.7% H_2 and from hemicellulose and lignin are approximately 25% CO, 36% CO_2, and 32% H_2. Pure cellulose produces a lower tar concentration in the syngas compared to beech and willow, which may be due to the hemicellulose and lignin content [6]. Study of the cell wall composition of the untreated and treated wood samples will aid in understanding the cause of treated wood property changes. The biochemical composition of biomass can be characterized using a thermogravimetric analyzer (TGA) [34,35]. Chen et al. qualitatively studied the composition of torrefied biomass (<150 μm, bamboo, banyan and willow) at 250, 275 and 300 °C using a TGA [28]. Mafu et al. quantitatively analyzed torrefied biomass (<500 μm (<0.500 mm), softwood chips, hardwood chips and sweet sorghum bagasse) at 260 °C using the 'food industry' chemical method [29]. For simplicity, this study utilized the TGA method.

Thermal pretreatments affect the biomass properties (such as shrinkage and grindablity) on a microscopic scale as well as on a macroscopic scale. Study of microstructural transformations will allow insight into the structural features and mechanisms responsible for the property changes. Microstructural transformations that occur during thermal degradation of the biomass are commonly observed using a scanning electron microscope (SEM) [36]. For this study, the same analysis locations in the three sample directions were followed during treatments so that change could be correlated to the different structures in each direction; a more detailed SEM analysis approach than typically reported [28,37]. The objectives of this study are to investigate a single wood particles over a range encompassing temperatures of torrefaction and low temperature pyrolysis; analyze the cell wall compositional changes for the treated woods using TGA; characterize physical and chemical property changes of the treated wood; and to examine the morphological changes of the wood samples during the thermal treatment process using SEM. These results improve the understanding of the property changes of the biomass during pretreatment and will help to develop models for process simulation and the treated biomass applications.

2. Experimental

This study focused on thermal treatment temperatures. Selected treatment temperatures were 220, 260, 300, 350, 450 and 550 °C. Selection of a wide range of temperatures was used to investigate both torrefaction and low temperature pyrolysis. The torrefaction temperatures of 220, 260, and 300 °C were selected to present light (200–235 °C), mild (235–275 °C), and severe (275–300 °C) torrefaction [15] and are considered temperatures reported in most studies [14,15]. The temperatures of pyrolysis were chosen based on reported studies of biochar fuel characteristics for direct combustion and co-firing with coal [10,12,19]. The pyrolysis temperature range is typically considered to be 350 to 650 °C [38]. After each of the thermal treatments, the samples were characterized and compared with untreated wood samples to evaluate property changes. The characterizations included biochemical composition/cell wall composition; physical (weight loss, color, dimensions/size, bulk density, and grindability) and chemical properties (proximate analysis and high heating value); and microstructual transformation.

2.1. Sample Preparation

Cherry wood (Prunus serotina; identification based on bark and wood), obtained locally in Pittsburgh, Pennsylvania, was selected as a model for this study. This relatively fast growing species is very common throughout the eastern United States and Canada and therefore has potential to be utilized as a biomass source. The wood samples used in this study were prepared by the following method (Figure 1). First, a 16-mm thick slice of wood was cut across the grain (longitudinal cell); second, an approximately 7 mm tangential section was split from the heartwood along an annual ring; and, finally, eighteen pieces having dimensions of about 16 mm × 7 mm × 5 mm (length, width, thickness) were split from the section. Samples were produced by splitting to avoid saw blade damage to the surfaces. The purpose of taking the samples from a single tangential section in the heartwood was to limit the effects of the natural variability of the wood and improve the reproducibility of results. For the eighteen samples (as illustrated in Figure 1), fourteen samples (1–7) (two sets) were used for biochemical composition/cell wall composition (1 set of 7), and proximate analysis and grindability (1 set of 7). Two samples (8–9) were used for the physical properties tests (weight, color, and size), and two samples (10–11) were used for the microstructure SEM analysis.

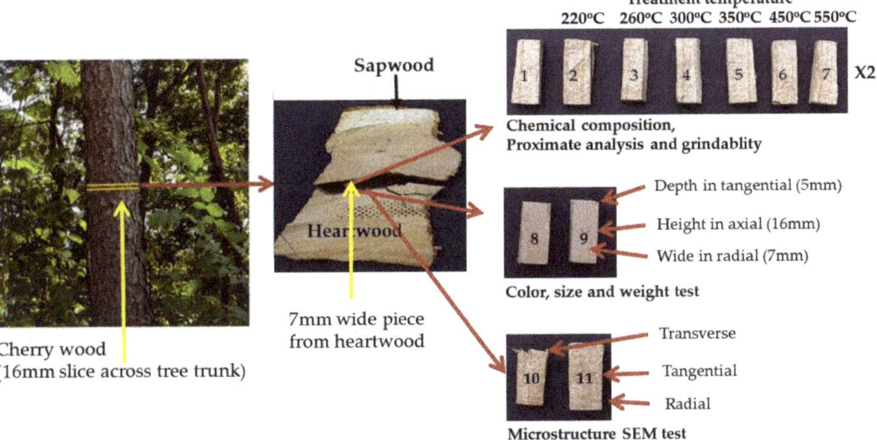

Figure 1. Sample preparation and sample arrangement.

2.2. Thermal Treatment

Figure 2 shows a schematic diagram of the thermal treatment experimental setup. The cherry wood samples were successively treated at 220, 260, 300, 350, 450 and 550 °C for 0.5 h for each temperature under flowing nitrogen in a quartz tube furnace. The samples were placed in a holder (Figure 2) and loaded into the middle of the tube at a fixed location and continuously purged with N_2 at a 50 mL/min flow rate for 0.5 h at room temperature to remove oxygen. Once the furnace reached the target temperature measured with a thermocouple, the quartz tube with the samples was placed in the furnace. When the samples reached the selected temperature, which was measured by a thermocouple placed close to the sample holder, they were held for 0.5 h. The treatment time was the same for all treatments since the treatment temperature is the more critical variable for torrefaction [15]. It was selected based on previously reported woody biomass torrefaction [22] and low temperature pyrolysis studies [12]. Torrefaction time is normally less than 1 h because the biomass thermal decomposition rate is initially fast, slowing after about 1 h [15]. A half hour was selected as the optimal time for torrefaction based on the improvement of grindability, the mass and energy yields [22]. A half hour treatment time was also used in a study of forest biomass grindability and fuel characteristics at torrefaction temperatures from 225 to 300 °C [20]. After the 0.5 h treatment, the tube was removed from the furnace and quickly cooled to room temperature under flowing N_2. The treated samples were stored in a desiccator.

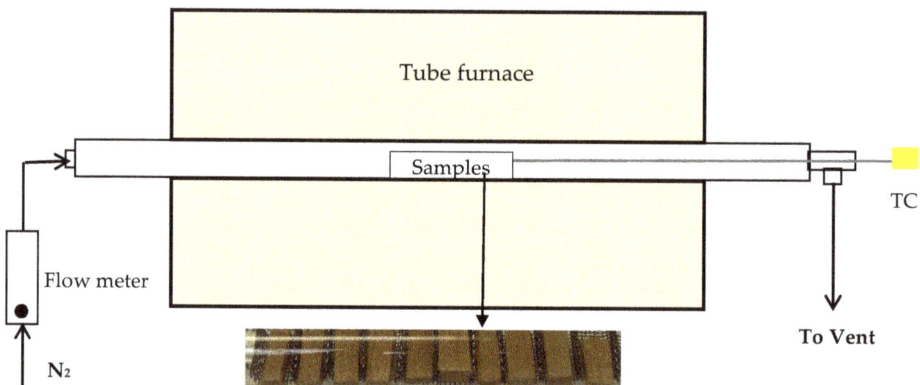

Figure 2. Schematic diagram of the thermal treatment experimental setup.

2.3. Characterization and Analysis of the Chars Generated from Thermal Reatments

After each treatment, the samples were characterized. Untreated wood (1, untreated 8–11) was used as a reference. For the twelve samples (2–7), one sample was removed after each temperature treatment in the sequence; therefore, each sample at a given temperature was cumulatively exposed to all the thermal treatments at lower temperatures. Two samples (8–9) were used to characterize physical property changes. The remaining two samples (10–11) were used for study of microstructural changes. These four samples (8–11) were taken through all the temperature treatments and examined after each step.

2.3.1. Cell Wall Component Decomposition Study of Untreated and Treated Woods by TGA and Differential Scanning Calorimetry (DSC)

For cell wall compositional changes, samples of approximately 10 mg were tested in the TGA (Perkin Elmer Pyris 1 TGA, PerkinElmer Inc., Shelton, CT, USA) using a non-isothermal method. Before the TGA test, the untreated and treated wood samples (1–7) were manually ground with a double-cut flat file and sieved through a 60 mesh screen (less than 250 μm). Samples were heated at a 2 °C/min from 25 to 800 °C under nitrogen with a total flow rate of approximately 125 mL/min. A slow heating rate was selected to clearly identify model components that thermally decompose over different temperature ranges. This rate was much lower than others reported (20 °C/min) [28,34].

Model compounds representing typical cell wall compositions were also analyzed by TGA to aid in identifying cell wall component changes in the treated wood samples. Cellulose (fibrous, long and medium), xylan from birch and beech (used as a model for hemicelluloses), and lignin (alkali) were selected as the model components. Both birch and beech are hardwoods and the use of xylans was to better represent hemicellulose because hemicellulose may be different between different types of biomass. All model compounds were purchased from Sigma-Aldrich (St. Louis, MO, USA).

Percent weight for each cell wall chemical component was calculated by weight loss in designated temperature ranges that correlate to the cell wall components from TGA analysis [34]. The designated temperature ranges were selected based on differential thermogravimetry (DTG) curves vs. temperature of the model cell wall components. Weight loss in each temperature range of the component was calculated by area in the range of DTG curve for the samples because the sample was heated at a slow constant rate and temperatures were linear with time. In this study, the area was adjusted based on the distribution of the model lignin in the designated temperature ranges because the lignin slowly decomposes over all temperature ranges (results shown in Section 3.1). Percent weight (dry basis) for the component was calculated by the weight loss in the temperature range of the component over the total weight loss in all designated temperature ranges (200–800 °C). The value of this method has been demonstrated with other biomass materials such as untreated and thermally treated switchgrass and cherry bark in our laboratory.

For differential scanning calorimetry (DSC) (Perkin Elmer Pyris Diamond DSC, PerkinElmer Inc., Shelton, CT, USA), the samples were tested at the same conditions as was used for TGA but over a temperature range of 25 to 600 °C. The DSC tests were used to further understand the thermal behaviors of biomass torrefaction and low temperature pyrolysis. All TGA and DSC experiments were performed in duplicate.

2.3.2. Physical Properties (Color, Weight, Dimensions, Bulk Density and Grindbility) of Untreated and Treated Woods

For physical changes, sample color was recorded using a digital camera and the dimensions measured using a digital caliper. After each treatment, the sample was weighed and weight loss percentage was calculated by the weight of treated sample over the untreated sample (mass yield = 100 − weight loss percentage). Dimensions of the samples are presented as height along the axial (longitudinal) direction, width along the radial direction and depth along the tangential direction (Figure 3) [39]. Bulk densities of the samples were calculated based on the measured weights and dimensions, however, due to the variations in sample shapes, the calculated bulk densities are approximate. The results of the bulk density measurement were used to estimate heating value in a volumetric basis. The physical property tests were performed in duplicate using samples (7–8). One sample was marked with a small "X" using a scalpel to differentiate it from the other.

Grindability testing of untreated and treated particles (1–7) was conducted using a hammer mill (Kinematica AG Polymix PX-MFC 90D, Luzernerstrasse, Switzerland) with a 0.8 mm bottom sieve at a rotational speed of about 1500 rpm. After milling, the samples were sieved using 500, 260 and 106 μm screen to four size fractions: >500 μm, 500–212 μm, 212–106 μm and <106 μm. Each fraction was weighted and its weight percentage was calculated over total sample weight to obtain particle size distribution for the sample. The fine particle size weight percentages of treated samples were compared with the untreated sample to evaluate the grindability changes resulting from treatment. The grindability test was performed one time for each treatment. However, multiple preliminary tests were conducted to determine the best grindability test method.

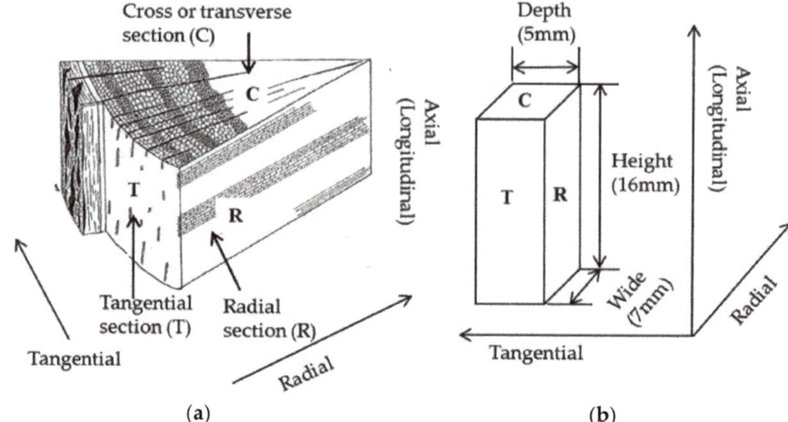

(a)   (b)

Figure 3. (a) Illustration of a section of wood showing the three planes discussed [39]; (b) Schematic of a wood sample used for these tests with the dimensions and orientations indicated.

2.3.3. Proximate Analysis and High Heating Value of Untreated and Treated Woods

For chemical property changes, proximate analysis (moisture content (MC), volatile matter (VM), fixed carbon (FC) and ash) was conducted by TGA (TA DSC-TGA Q600). The ASTM D-3172 (ASTM International, West Conshohocken, PA, USA) procedure for coal and coke was used. The samples used for the proximate analysis were from the grindablity tests. Approximately, 10 mg of ground sample with a size of <212 μm was placed in an alumina crucible for testing. MC and VM were tested at 105 and 950 °C, respectively, in Ar. Ashing was conducted at 750 °C in 22% O_2 with Ar balance. FC was calculated on a dry basis by subtracting ash and VM percentages. The proximate analyses were performed in duplicate.

High heating value (HHV) was calculated using proximate analyses and cell wall composition (structure analysis) data. This is simple method for the estimation of caloric value and is acceptable for engineering estimations of net heating value of a biomass fuel [40]. In this paper, the correlation proposed based on structural analysis (lignin content) (Equation (1)) [41] and the proximate analysis (Equation (2)) [42] were selected. Those correlations developed based on raw biomass are also applicable to the estimation of the HHV of torrefied biomass [15].

For wood,

$$HHV = 0.0893 \, (L) + 16.9742 \tag{1}$$

where HHV (MJ/kg) is the energy content on a dry-ash-free and extractive-free basis. L (wt %) is lignin content of biomass on a dry basis.

$$HHV = 0.3536FC + 0.1559VM - 0.0078Ash \tag{2}$$

where HHV (MJ/kg) is the energy content on a dry basis. FC (wt %), VM (wt %) and ash (wt %) are fixed carbon, volatile matter and ash on a dry basis, respectively.

2.3.4. Microstructual Transformations Study Using SEM

For the SEM analyses (FEI Company (Thermo Fisher Scientific, Waltham, Massachusetts, USA) Quanta 600 field emission scanning electron microscope, low vacuum mode, secondary electron detector), three plane sections on the wood sample (tangential, radial, and cross or transverse) and twelve analysis spots for each section were examined using six magnifications at each spot resulting in a total of 216 images. The tangential plane section of the wood sample (10) (with the rough edge, indicated by the arrow in Figure 1) was prepared by manually breaking it to preserve its microstructure. The sample (11) was used for the microstructure study of the radial and transverse plane sections of the wood sample (indicated by the arrow in Figure 1). To allow the same analysis sites to be followed through all the treatment steps, small reference marks were made on the samples with a scalpel. Low magnification images were collected and a reference map was constructed allowing the reference marks and obvious features to be utilized for the relocation of the analysis sites following each treatment step. Following the changes observed at the same analysis spots through the thermal treatment steps allows correlation of the observed changes with other measured characteristics. These results contribute to the understanding of the property changes of the biomass during treatment.

3. Results and Discussions

3.1. Cell Wall Component Composition Study of Untreated and Treated Woods by TGA and DSC

TGA analysis of the decomposition of pure cellulose, xylan and lignin was used to identify the temperature ranges that will be used for the cell well composition analysis of untreated and treated woods. Figure 4 shows the thermogravimetry (TG) and the differential thermogravimetry (DTG) curves vs. temperature for long and medium cellulose fibers, xylans from beech and birch wood (representing hemicelluloses) and lignin (alkali). The DTG curves show a distinct primary peak for each model component. Xylan from birchwood and xylan from beechwood decomposed over the range of about 200 to 300 °C with peak maximums at 269 and 274 °C, respectively. The long and medium cellulose fibers mainly decomposed over the range of about 300 to 360 °C with peak maximums at 330 and 328 °C, respectively. Lignin slowly decomposed over the range of about 200 to 800 °C with a peak maximum at 353 °C. The hemicelluloses are the most thermally sensitive of the main biomass components. Yang et al. studied pyrolysis of hemicellulose (xylan from birchwood), cellulose fiber and lignin (alkali) and reported that xylan decomposed at 220–315 °C with peak maximums at 268 °C, cellulose fiber at 315–400 °C with peak maximums at 355 °C, and lignin at 150–900 °C [43]. Our results showed a lower decomposition temperature due to our lower heating rate of 2 °C/min versus their 10 °C/min. The peak maximums of celluloses measured by both TGA and DSC were comparable. The peak maximums observed for the xylans and lignin by TGA were slightly lower than those measured by DSC but still fell in similar decomposition temperature ranges.

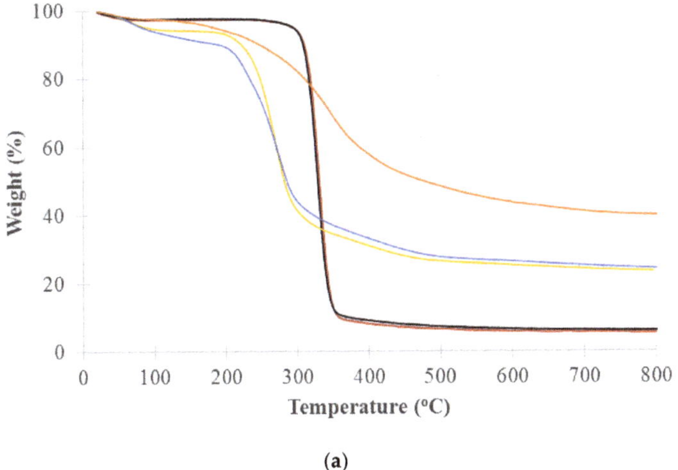

(a)

Figure 4. *Cont.*

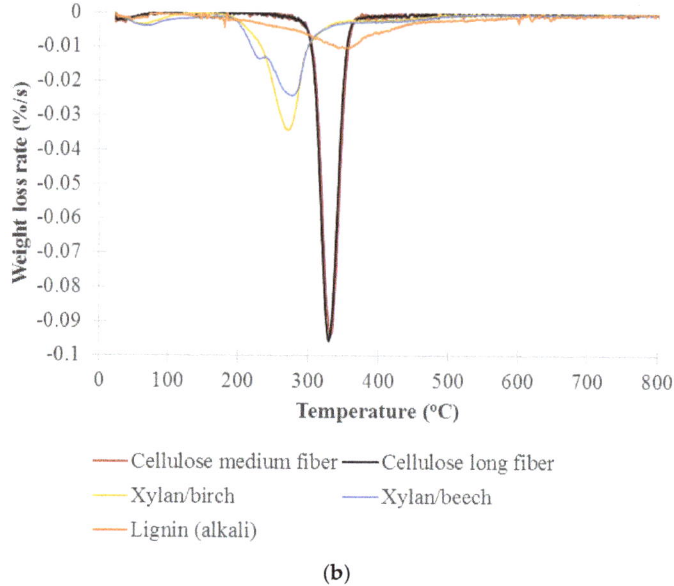

(b)

Figure 4. (a) Thermogravimetry (TG); (b) Differential thermogravimetry (DTG) curves versus temperature for the pure model components (xylan/hemicelluloses, cellulose, and lignin) of biomass cell wall components.

Percent weight for each cell wall chemical component was calculated by weight loss in designated temperature ranges that correlate to the cell wall components from TGA analysis [34]. Based on DTG curves of the model components (Figure 4b), the designated temperature ranges for calculating the percent weight of cell wall chemical components were selected. The ranges are hemicellulose between 200 and 300 °C, cellulose between 300 and 360 °C, and lignin between 360 and 800 °C. Percent weight (dry basis) for the component was calculated by the weight loss in the temperature range of the component over the total weight loss in all designated temperature ranges (200–800 °C). The weight loss in each temperature range was calculated by area in the range of DTG curve for the samples. In this study, the area was adjusted based on the distribution of the model lignin in the designated temperature ranges. Table 1 lists the calculated cell wall chemical components of the untreated and treated woods at 220, 260, 300, 350, 450 and 550 °C. The results clearly show the changes in cell wall chemical components with thermal treatment temperature. Hemicellulose contents of the untreated and torrefied woods at 220 , 260 ,and 300 °C are 24.3% 21.2%, 10.1% and 7.4%, respectively, showing that hemicellulose was slightly decomposed by 220 °C, mildly by 260 °C, and significantly by 300 °C. Cellulose was slightly decomposed for the torrefied wood at 300 °C and more severely for the pyrolyzed wood at 350 °C. The samples pyrolyzed at 450 and 550 °C have only lignin with both hemicellulose and cellulose decomposed. The cell wall composition of the torrefied woods correspond with the light, mild and severe torrefaction as classified by Chen et al. [15] and are similar with the results reported by Chen et al. [28] and Mafu et al. [29]. Mafu et al. tested the hardwood cell wall compositions of untreated and torrefied at 260 °C samples using the chemical method. Hemicellulose, cellulose and lignin contents of the untreated and torrefied samples were 11.4% and 1.2%, 56.9% and 46.7%, 15.7% and 16.0%, respectively. Most of hemicellulose for the torrefied wood was decomposed while cellulose and lignin were slightly impacted or not changed.

Table 1. Calculated cell wall chemical component of untreated and treated woods at each treatment temperature based on thermogravimetric analyzer (TGA) tests.

Samples	Chemical Components (%)		
	Hemicellulose	Cellulose	Lignin
Untreated	24.3	42.8	32.9
220	21.2	46.1	32.8
260	10.1	48.3	41.6
300	7.4	38.7	53.9
350	0.0	6.8	93.2
450	0.0	0.0	100.0
550	0.0	0.0	100.0

The decomposition of cellulose, xylan, and lignin were endothermic in the temperature range up to 600 °C as shown by the DSC results. The order of the amount of energy (heat of decomposition) required from high to low is: Cellulose > xylan > lignin. Decomposition temperature ranges and peak temperatures for cellulose, xylan and lignin obtained from DSC tests are comparable to TGA. During the DSC tests, the amount of energy absorbed (heat of decomposition) for treated samples at 350, 450 and 550 °C was much less than observed for the lower temperature treated woods at 220, 260 and 300 °C since at 350 °C and above, the xylans/hemicelluloses and much of the cellulose, which has the highest heat of decomposition of all the cell wall components of biomass, have already been decomposed. For the sample treated at 220 °C, some hemicellulose is decomposed. For the samples treated at 260 and 300 °C, hemicellulose is mostly decomposed. Both hemicellulose and cellulose are decomposed for the samples treated at 450 and 550 °C.

Decomposition of the cell wall components of the cherry wood during the thermal treatments can be used to explain the cause of the grindability improvement after the thermal treatments (grindability test results in Section 3.2). Wood strength results primarily from the fiber content since wood is basically a series of tubular fibers or cells cemented together [32,44]. The cell walls have three main regions: the middle lamella (ML) that acts as adhesion between two or more cells, the primary wall (P) and the secondary wall in its three layers (S1, S2,and S3) (Figure 5a) [44]. These regions have different thicknesses and various quantities of the major chemical components hemicellulose, cellulose, and lignin. For example, for Scotch pine wood, the S2 layer has the highest cellulose content (32.7%) in the cell wall with lesser quantities of hemicellulose (18.4%) and lignin (9.1%) [32]. The S2 layer is the thickest in the secondary wall and in the cell wall overall. From this respect, the S2 layer is mainly responsible for the overall properties of the cell wall. For the S2 layer, a proposed ultrastructural model described the arrangements of hemicellulose, cellulose and lignin and is illustrated in Figure 5b [32]. Lignin and hemicellulose act as a matrix for cellulose in the S2 layer. The lignin in the S2 layer is evenly distributed throughout the layer. Thin, low molecular weight hemicellulose channels bond cellulose microfibrils on their radial faces. Sheaths of hemicellulose are along the boundary area of the cellulose microfibrils. Thermal treatments decompose the hemicellulose resulting in interwall cracks in the S2 layer and weakening of the cell wall. The resulting weakening leads to improved biomass grindability. For the samples treated at 260 and 300 °C, hemicellulose was mostly decomposed resulting in a weakening of the cell walls and, subsequently, an improvement in grindability.

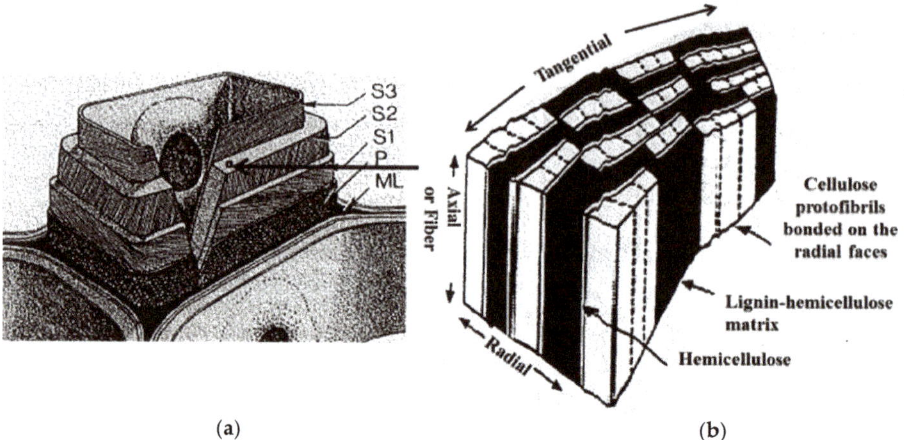

(a)   (b)

Figure 5. (**a**) Illustration of three main regions in the cell wall: the middle lamella (ML), the primary wall (P), and the three layers of the secondary wall (S1, S2 and S3) [44]; (**b**) Proposed ultrastructural model of the arrangement of hemicellulose, cellulose and lignin in the S2 layer [32].

Torrefaction decomposes most of the hemicellulose. Therefore, gasification of torrefied wood should result in improve syngas quality since hemicellulose and lignin may produces a higher tar concentration in the syngas compared to cellulose [6]. This benefit is also confirmed by gasification of torrefied biomass pellets which resulted in both improved energy efficiency and syngas quality [15,23]. The lignin contents of treated woods increased as the treatment temperature increased (Table 1). The high heating values of lignin are higher than the cellulose and hemicellulose due to higher degree of oxidation of the cellulose and hemicellulose [4]. The biochar with higher lignin content has the high heating values that is desired combustion property. On other hand, the air-steam gasification conversions of lignin (52.8%) was much lower than cellulose (97.9%) and hemicellulose (92.2%) [34]. Therefore, the production method of biochar used for gasification should be considered.

3.2. Physical Properties of Untreated and Treated Woods

3.2.1. Physic Properties of Color, Weight, Dimensions, and Bulk Density

Table 2 shows the physical properties of untreated versus treated wood at 220, 260, 300, 350, 450 and 550 °C. Thermal treatments were found to darken the wood samples, as expected. The samples changed from the light brown of the untreated wood to dark cinnamon after the 220 °C treatment and essentially black after the 260 °C treatment. The darkening of the wood might be due to losses of arabinose and xylose that are converted to chocolate-brown colored furfural [32]. This agreed with the TGA test results on chemical composition changes of treated woods. Some hemicellulose in the 220 °C treated wood, most hemicellulose in the samples treated at 260 and 300 °C, and all hemicellulose in the samples treated at 350, 450 and 550 °C was decomposed.

The dimensions of the wood were reduced in all directions and shrinkage increased as treatment temperature increased (Table 2). The shrinkage observed is a result of the chemical reactions associated with cell wall component decomposition and subsequent structural changes during the thermal treatments. In the axial/longitudinal direction (height), the shrinkage for the 220 and 260 °C treated samples were close to zero; shrinkage started at 300 °C (0.41%) and increased significantly at 350 °C (6.48%). Our results are similar to Davidsson and Pettersson who reported that the longitudinal shrinkage of birch wood particles was close to zero below 300 °C and 5% at 350 °C. The longitudinal shrinkage of wood is a result of decomposition of cellulose in the microfibrils in the S2 layer of the cell walls according to the microfibril dominance theory suggested [45]. Our TGA test results agreed with the axial shrinkage. Cellulose in the 220 °C and 260 °C treated woods did not decompose during the thermal treatment. Some cellulose in the 300 °C treated wood decomposed and cellulose in the 350 °C treated wood decomposed completely during these thermal treatments.

Table 2. Physical property changes of untreated and treated woods at each treatment temperature.

Physical Property	Temperature (°C)						
	25	220	260	300	350	450	550
Color							
Weight loss (%)	0	15.84	27.57	39.21	62.07	69.70	73.22
Height shrinkage (%)	0	0.09	0.06	0.41	6.48	12.51	15.34
Wide shrinkage (%)	0	3.85	6.64	10.72	22.54	26.95	29.53
Depth shrinkage (%)	0	1.06	2.31	4.52	13.57	19.44	22.24
Volume loss (%)	0	4.96	8.85	15.11	37.39	48.52	53.60
Bulk density (g/cm^3)	0.54	0.48	0.43	0.39	0.33	0.32	0.31
Density change (%)	0	−11.44	−20.54	−28.39	−39.43	−41.16	−42.29

The shrinkages along radial (width) and tangential directions (depth) were similar in behavior and were different from the axial shrinkage. The shrinkages in these two directions started at low temperature, approximately 220 °C, primarily due to hemicellulose decomposition (Table 1). A large increase in the percent shrinkage was observed between 300 and 350 °C due to cellulose decomposition. Both the radial and tangential direction shrinkages were more than the axial direction. This may be because decomposition of hemicellulose and lignin mainly contributed to the radial and tangential shrinkages. Since cellulose is arranged in fibrils mainly in the axial directions and both hemicellulose and lignin bond the cellulose fibrils together, hemicellulose decomposition, primarily at low temperature, and lignin decomposition, in all temperature ranges, with a peak at a temperature higher than the cellulose decomposition temperature, is primarily responsible for the observed shrinkage (Figure 4). The order of shrinkages from high to low was width > depth > height. Based on the ultrastructural model (Figure 5b), the thin hemicellulose channels bond cellulose microfibrils on their radial faces and the hemicellulose decompositions results in the shrinkage in the radial direction.

The shrinkage was caused by the decomposition of cell wall components that resulted in loss of a portion of the mass from the biomass. Regression analyses of the shrinkages (S%) along radial, tangential and longitudinal direction as a function of the weight loss (w %) during the thermal pretreatments were carried out on experimental data. Formulas and R-squared (R^2) of the regression analysis are listed in Equations (3)–(5). The formulas show good agreement with experimental data.

For radial direction/width,

$$Sr = 2.4533e^{0.0349w}, \ R^2 = 0.9911 \tag{3}$$

For tangential direction/depth,

$$St = 0.5201e^{0.0522w}, \ R^2 = 0.9955 \tag{4}$$

For tangential direction/Height,

$$Sl = 0.0085e^{0.103w}, \ R^2 = 0.9538 \tag{5}$$

The variations in shrinkage in the three sample reference directions are important for modeling studies of the wood thermal pretreatment. The shrinkage affects heat transfer to the particle and gas flow within the particle [25].

The weight losses of treated samples at 260 and 300 °C were 27.57% and 39.21%, respectively (Table 2). It was mainly caused by hemicellulose decomposition/devatilization to volatiles along with the cellulose and lignin decomposition (Table 1). These results were comparable to Yan et al. who reported that mass yields (100 - weight loss%) of pine woods after torrefaction at 275 and 300 °C were 74.2% and 60.5%, respectively [46]. Losses of weight and volume increased as treatment temperature increased. Losses of weight and volume of the pyrolyzed wood samples were much higher than those of the torrefied samples. There was a large step in percentage weight loss between 300 and 350 °C, which is more than for any other temperature step. This is due to the decomposition of cellulose which the primary component of the wood. At 350 °C, the weight loss (62.07%) and volume loss (37.39 %) were much higher than the weight loss (39.21 %) and volume loss (15.11%) at 300 °C. The weight losses at the higher temperature treatments of 450 to 550 °C increased slightly from 69.70 to 73.22%. Those results were close to Abdullah and Wu who reported that char yields of treated mallee wood at 300 and 450 °C were ~56% and ~27%, respectively [10]. The weight loss during low temperature pyrolysis was much higher than observed during torrefaction.

The measured bulk density of the untreated cherry wood was 540 kg/m^3 (0.54 g/cm^3) which falls in the typical range [16], The densities of the treated wood samples decreased with increasing temperature over the temperature range of 220 to 350 °C but changed little over the range of 350 to 550 °C. The bulk density of wood is dependent upon cellular diameters and wall thickness [47]. During the thermal treatments, weight loss and size reduction caused by the decomposition of chemical components in the cell wall resulted in the density changes. Weight losses higher than volume losses resulted in the density reduction with increasing temperatures. The low density of treated wood may be addressed through densification (pelletizing for example) to improve the handling, transportation, storage and conversion of the wood [14,15].

Regression analyses of the weight loss (w%), volume loss (v%) and density (d g/cm3) as a function of the treatment temperature (T °C) were carried out on experimental data. Formulas and R-squared (R^2) of regression were listed in Equations (6)–(8). The formulas show good agreement with experimental data.

For weight loss,

$$w = -0.0007T^2 + 0.7451T - 114.31, R^2 = 0.9788 \tag{6}$$

For volume loss,

$$v = -0.0004T^2 + 0.4607T - 81.733, R^2 = 0.9555 \tag{7}$$

For density,

$$d = 0.000003T^2 - 0.00261T + 0.9146, R^2 = 0.9774 \tag{8}$$

3.2.2. Grindbility of Untreated and Treated Woods

The untreated and treated at 220 °C samples were incompletely milled resulting in a fraction of those materials not being able to pass the bottom screen (0.8 mm) of the hammer mill. The untreated wood was ground three times and during grinding it was occasionally stuck in the mill between hammer-grinding stators and hammer-rotors. The ground samples were sieved into four size fractions: >500 μm, 500–212 μm, 212–106 μm and <106 μm. Table 3 lists the particle size distribution of the milled untreated and treated woods. The grindability of treated woods improved as the percentage of particles with the lower size fractions increased. The treated wood at 260 and 300 °C had 23.5% and 55.5% passing through 212 μm, respectively, compared to only 11% of the untreated wood. The results are similar to Aria et al. who reported that the percentage of particles less than 150 μm was double that of untreated wood for wood treated at 240 °C for 0.5 h and ground.

From the cell well composition study (Table 1), hemicellulose contents of the treated woods at 260 °C (10%) and 300 °C (7%) were lower that of the untreated wood (24%). Wood is essentially a series of elongated tubular fibers or cells and cemented together [31,32]. Hemicellulose acts as a matrix for the cellulose along with lignin. The decomposed hemicellulose resulted in the cell walls weakening and subsequently improved grindability as shown by the increase in the percentage of small particles versus the untreated wood.

For the treated wood at 350 °C, the percentage that passed through the 106 μm sieve increased to 16.7% compared to 3% of the treated wood at 300 °C. However, further treatment at higher temperatures did not further improve the grindability. The results are similar to Abdullah and Wu (2009) who reported that grindability of treated wood at 300 °C was drastically improved but increasing temperature to 500 °C resulted in only a small additional improvement in the grindability [10]. Therefore the torrefaction of wood can improve the wood grindability. The torrefied wood may be directly fed with coal for co-firing in a PC boiler and for co-gasification in an entrained-flow gasifier that require fine particle size of fuels.

Table 3. Particle size distribution of the milled untreated and treated woods.

Samples	>500 μm	500–212 μm	212–106 μm	<106 μm
Untreated	66.0	40.7	11.0	2.8
220	63.6	45.9	11.9	1.5
260	54.2	53.2	23.5	3.6
300	32.3	38.1	55.5	3.0
350	4.2	33.6	45.5	16.7
450	4.6	23.6	50.5	21.4
550	3.8	24.7	49.8	21.8

3.3. Proximate Analysis and High Heating Value of Untreated and Treated Woods

The proximate analysis of untreated and treated woods reveals the changes in moisture, volatile matter, fixed carbon and ash after thermal pretreatment. The moisture content of the untreated wood sample used in this study was low at 6% and similar to the 6.5% of reported for oak wood [16]. Table 3 shows the proximate analysis results of untreated and treated woods at 220, 260, 300, 350, 450 and 550 °C. The VM of untreated wood is high (88.1%), while its FC (11.5%) and ash (0.4%) are low. These untreated wood proximate analysis results are similar to that reported for beech wood with VM 84.2%, FC 15.5% and ash 0.3% [14]. The thermally treated woods were decomposed and light volatiles lost so the VM in the treated woods decreased and the FC increased with increasing temperature. The FC of pyrolyzed woods at 350, 450 and 550 °C are 54.3 to 83.4% and much higher than those of the torrefied woods at 220, 260 and 300 °C 13.9 to 24.9%. The pyrolyzed woods has high fuel quality, similar to coal which has mean FC 43.9% [48].

The calculated HHVs of untreated and treated woods in weight (MJ/kg) and volumetric value (MJ/m^3) were listed in Table 4. The HHVs of torrefied woods at 220, 260 and 300 °C were 18 to 22 MJ/kg and in the range of 16–29 MJ/kg of torrefied biomass [15]. The heating values (by weight) of torrefied woods at 260 and 300 °C increased 6.1% and 15.9% compared to the untreated wood. The torrefied wood with improved fuel quality may be utilized as solid fuel with higher process efficiency. The pyrolyzed woods at 350, 450 and 550 °C were 25 to 32 MJ/kg and close to the mean HHV of coal of 25 MJ/kg [48]. Therefore, the pyrolyzed woods are compatible for utilization with coal. The HHVs of the treated woods by weight increased with increasing treated temperature due to increasing weight loss due to devolatilization hemicellulose and cellulose. But the HHVs of the treated woods in a volumetric basis decreased because weight losses were higher than volume losses. These results agree with those reported [17]. Pelletizing can significantly increase the volumetric energy density of the treated biomass facilitating transport and storage, leading to savings in logistics [9].

Fuel ratio is defined as a ratio of fixed carbon to volatile matter (FC/VM) and is an important solid fuel property [5]. From the proximate analysis results (Table 4), fuel ratio of the treated woods increased as the volatiles decreased due to the thermal treatment. The fuel ratios of the torrefied woods were 0.16–0.34 and lower than lignite coal 0.85 [5]. Those of the biochars at 350, 450 and 550 °C were 1.2, 2.8 and 5.6, respectively. The biochar from the wood treated at 350 °C (1.2) was close to that of high volatile bituminous coal (0.92) [19] and bituminous coal 1.56 [20]. So the low temperature pyrolysis improves wood combustion properties but the torrefaction did not [12]. The biochars were more suitable than raw biomas for co-firing with coal and increased thermal conversion efficiency [5,24].

Table 4. Proximate analysis and high heating value changes of untreated and treated woods at each treatment temperature.

Samples	VM	FC	Ash	High Heating Value (MJ/kg)				High Heating Value (MJ/m^3)			
				Proximate	Incr.	Lignine	Incr.	Proximate	Decr.	Lignine	Decr.
	(wt%)	(wt%)	(wt%)	Eq. 2	%	Eq. 1	%	Eq. 2	%	Eq. 1	%
Untreated	88.1	11.5	0.4	17.8		19.7		9605.2		10663.2	
220	85.7	13.9	0.5	18.3	2.6	19.7	0.0	8760.0	−8.8	9474.1	−11.2
260	82.0	16.0	0.8	18.9	6.1	20.5	3.9	8113.0	−15.5	8823.6	−17.3
300	73.3	24.9	0.7	20.6	15.9	21.6	9.5	8041.5	−16.3	8429.3	−20.9
350	44.1	54.3	0.9	26.3	47.9	25.1	27.1	8682.0	−9.6	8285.4	−22.3
450	25.8	72.9	1.6	29.7	66.9	25.7	30.2	9500.3	−1.1	8227.8	−22.8
550	14.9	83.4	1.7	31.8	78.7	–	–	9856.0	2.6	–	–

During the thermal treatment, the mass loss results in energy loss with respect to the untreated biomass. The treated biomass is a solid fuel so its energy content can be evaluated using energy yield. The energy yield is defined as: energy yield (%) = mass yield (%) * HHVf/HHVo, where mass yield = 100 − weight loss (%), HHV is high heating value, subscript o and f refer to the untreated and treated biomass, respectively [22]. The energy yields of the torrefied woods were in the range 70.3–86.5% and those of the pyrolyzed woods in the range 41.8–56.0%. With respect to energy yield, the torrefied woods are superior compared to the pyrolyzed woods [12]. The energy loss may be recovered by utilizing the volatiles generated during the pretreatment.

3.4. Morphological Study of Untreated and Treated Woods by SEM

Figure 6 shows the SEM images (horizontal field width = 1.28 mm; scale bar of 500 μm; vertical as images are presented) of the tangential section views of the untreated and the 260 and 550 °C treated woods. There are fibers, rays seen in end-view formed by stacks of ray parenchyma cells and pores visible in the tangential section images. Identified morphological structures were seen in all three woods. Figure 7 shows the SEM images (horizontal field width = 1.28 mm; scale bar of 500 μm; vertical as images are presented) of the radial section views of the untreated and the 300 and 550 °C treated woods. There are fibers, rays that look like a brick wall crossing in the longitudinal direction and pores visible in the radial section images. Figure 8 shows the SEM images (horizontal field width = 1.28 mm; scale bar of 500 μm) of the cross section views of the untreated and the 300 and 550 °C treated woods. There are many pores of various sizes and arrangements, fibers and annual or growth rings visible in the cross section images. Overall, the morphological features of the wood remain intact during the thermal treatments, which is in agreement with Haas et al. [49] who studied the pyrolysis of poplar wood up to 700 °C using microscopy in real-time and Winandy and Rowell [32] who studied the pyrolysis of pine wood at 295 °C. Figure 9 shows the SEM images (horizontal field width = 320 μm; scale bar of 100 μm) of cross sections of the untreated and the 350 and 450 °C treated samples. The cell walls were thinner after the thermal treatments.

Along the axial direction (Figure 6), the untreated and the 260 °C treated samples did not significantly shrink: however, the 550 °C treated sample shrank about 15%, which agreed with the physical property tests of sample height (Table 2). Along the tangential direction, the 260 °C treated samples shrank slightly; however, the 550 °C treated samples shrank 22% compared with the untreated sample, which agreed with the physical property tests of sample depth (Table 2). There was more shrinkage along the tangential direction than the axial direction. Along the radial direction (Figure 7), the samples treated at 300 and 550 °C shrank 10 and 29% respectively, compared with the untreated one, which agreed with physical property test of sample width, Table 1.

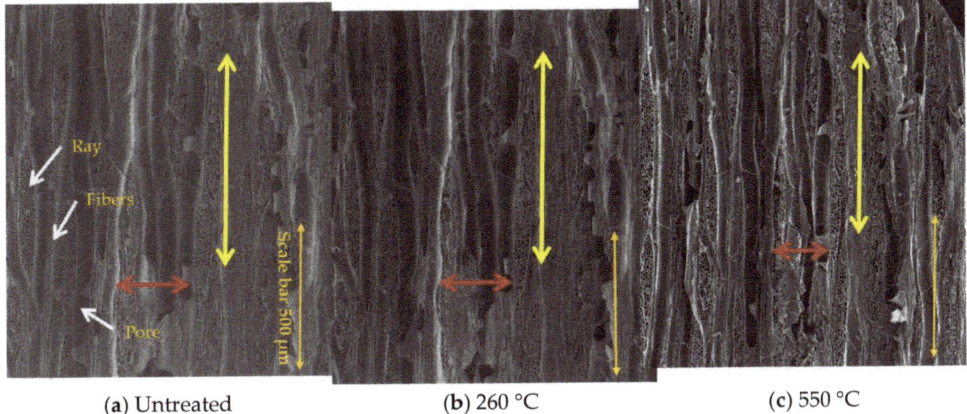

Figure 6. SEM images (horizontal field width = 1.28 mm; scale bar of 500 μm; vertical as images are presented) of the tangential section views of the untreated and the 260 and 550 °C treated cherry wood samples. White arrows indicate rays and fibers seen in end-view formed by stacks of ray parenchyma cells, also pores are visible in the tangential section images. Yellow bars indicate shrinkage along the axial direction using reference structures. Red bars mark shrinkage along the tangential direction using reference structures.

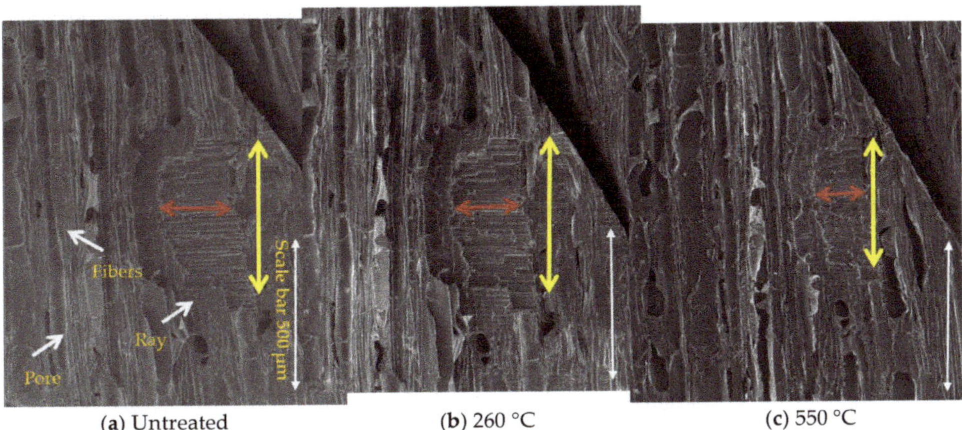

Figure 7. SEM images (horizontal field width = 1.28 mm; scale bar of 500 μm; vertical as images are presented) of the radial section views of the untreated and the 300 and 550 °C treated cherry wood samples. White arrows point out fibers and rays that look like a brick wall crossing in the longitudinal direction, also pores are visible in the radial section images. Yellow bars mark shrinkage along the axial direction (using reference structures). Red bars mark shrinkage along the radial direction (using reference structures).

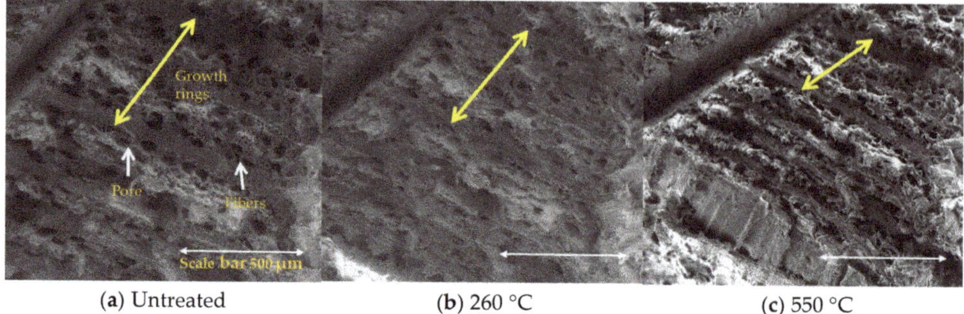

(**a**) Untreated   (**b**) 260 °C   (**c**) 550 °C

Figure 8. SEM images (horizontal field width = 1.28 mm; scale bar of 500 μm) of the cross section views of the untreated and the 300 and 550 °C treated cherry wood samples. White arrows point out fibers and pores. Yellow bars mark annual or growth rings and its shrinkage along the radial direction (using reference structures).

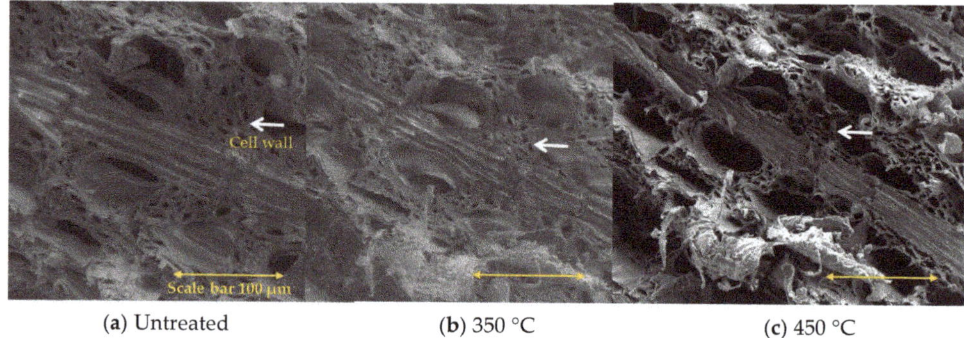

(**a**) Untreated   (**b**) 350 °C   (**c**) 450 °C

Figure 9. SEM images (horizontal field width = 320 μm; scale bar of 100 μm) of the cross section views of the untreated and the 350 and 450 °C treated cherry wood samples. White bars point out cell walls which were thinner after the thermal treatments.

4. Conclusions

This study utilized well defined wood particle samples to obtain torrefaction and low temperature pyrolysis data to more thoroughly understand the processes and aid future modeling studies. Treatment temperature has a great impact on biochemical composition, physical and chemical properties and microstructure of wood during torrefaction (220–300 °C) and low temperature pyrolysis (300–550 °C). The hemicellulose in the wood treated at 260 and 300 °C was mostly decomposed and resulted in the cell walls weakening and subsequent improvement in grindability as shown by the increase in the percentage of small particles as compared to the untreated wood. Treatment at higher temperatures (>300 °C) did not result in a significant increased improvement of grindability. The morphologies of the wood samples remained essentially intact over the treatment temperature range but the cell walls were observed to be thinner. Shrinkages of the treated woods in the three reference directions increased with increased treatment temperature and correlated well with each other. Losses of weight and volume of the pyrolyzed wood samples were much higher than those of the torrefied samples. The bulk densities of the treated wood samples decreased with increasing temperature over the temperature range of 220 to 350 °C but changed little over the range of 350 to 550 °C. The HHVs of the treated woods by weight increased with increased treatment temperature but the HHVs of the treated woods in a volumetric basis decreased. The low temperature pyrolyzed wood samples resulted in an improved solid fuel high fuel ratio, which was close to that of lignite/bituminous coal. The selection of a biomass thermal pretreatment process should take into consideration the application because torrefaction and low temperature pyrolysis result in different product properties. These results improve the understanding of the property changes of the biomass during thermal pretreatment and will help in the development of models for process simulation and potential application of the treated biomass.

Acknowledgments: This report was prepared as an account of work sponsored by the Department of Energy, National Energy Technology Laboratory, an agency of the United States Government. Neither the United States Government nor any agency thereof, nor any of their employees, makes any warranty, express or implied, or assumes any legal liability or responsibility for the accuracy, completeness, or usefulness of any information, apparatus, product, or process disclosed, or represents that its use would not infringe privately owned rights. Reference herein to any specific commercial product, process, or service by trade name, trademark, manufacturer, or otherwise does not necessarily constitute or imply its endorsement, recommendation, or favoring by the United States Government or any agency thereof. The views and opinions of authors expressed herein do not necessarily state or reflect those of the United States Government or any agency thereof.

Author Contributions: The authors equally contributed to the work reported.

Conflicts of Interest: The authors declare no conflict of interest.

References

1. International Energy Agency (IEA). *Technology Roadmap Bioenergy for Heat and Power*; International Energy Agency: Paris, France, 2012; Available online: http://www.iea.org/publications/freepublications/publication/2012_Bioenergy_Roadmap_2nd_Edition_WEB.pdf (accessed on 10 May 2017).
2. U.S. Energy Information Administration (EIA). *International Energy Outlook 2016*; DOE/EIA-0484 (2016); U.S. Energy Information Administration: Washington, DC, USA, 2016. Available online: https://www.eia.gov/outlooks/ieo/ (accessed on 11 May 2017).
3. U.S. Energy Information Administration (EIA). *June 2017 Monthly Energy Review*; DOE/EIA-0035(2017/6); U.S. Energy Information Administration: Washington, DC, USA, 2017. Available online: https://www.eia.gov/totalenergy/data/monthly/pdf/mer.pdf (accessed on 11 May 2017).
4. Saidur, R.; Abdelaziz, E.A.; Demirbas, A.; Hossain, M.S.; Mekhilef, S. A review on biomass as a fuel for boilers. *Renew. Sustain. Energy Rev.* **2011**, *15*, 2262–2289. [CrossRef]
5. Liu, Z.; Balasubramanian, R. A comparison of thermal behaviors of raw biomass, pyrolytic biochar and their blends with lignite. *Bioresour. Technol.* **2013**, *146*, 371–378. [CrossRef] [PubMed]
6. Van Paasen, R.H.; Kiel, J.H.A. *Tar Formation in a Fluidised-Bed Gasifier: Impact of Fuel Properties and Operating Conditions*; Energy Research Center of the Netherlands (ECN): Sint Maartensvlotbrug, The Netherlands, 2004.
7. Higman, C.; van der Burgt, M. *Gasification*; Elsevier: New York, NY, USA, 2008.
8. Smoot, L.D.; Smith, P.J. *Coal Combustion and Gasification*; Plenum Press: New York, NY, USA, 1985.
9. Bergman, P.; Kiel, J. Torrefaction for biomass upgrading. In Proceedings of the 14th European Biomass Conference & Exhibition, Paris, France, 17–21 October 2005.
10. Abdullah, H.; Wu, H. Biochar as a fuel: 1. Properties and grindability of biochars produced from the pyrolysis of mallee wood under slow-heating conditions. *Energy Fuels* **2009**, *23*, 4174–4181. [CrossRef]
11. Huang, Y.-F.; Syu, F.-S.; Chiueh, P.-T.; Lo, S.-L. Life cycle assessment of biochar cofiring with coal. *Bioresour. Technol.* **2013**, *131*, 166–171. [CrossRef] [PubMed]
12. Park, S.-W.; Jang, C.-H.; Baek, K.-R.; Yang, J.-K. Torrefaction and low-temperature carbonization of woody biomass: Evaluation of fuel characteristics of the products. *Energy* **2012**, *45*, 676–685. [CrossRef]
13. Van der Stelt, M.J.C.; Gerhauser, H.; Kiel, J.H.A.; Ptasinski, K.J. Biomass upgrading by torrefaction for the production of biofuels: A review. *Biomass Bioenergy* **2011**, *35*, 3748–3762. [CrossRef]
14. Chew, J.J.; Doshi, V. Recent advances in biomass pretreatment—Torrefaction fundamentals and technology. *Renew. Sustain. Energy Rev.* **2011**, *15*, 4212–4222. [CrossRef]
15. Chen, W.-H.; Peng, J.; Bi, X.T. A state-of-the-art review of biomass torrefaction, densification and applications. *Renew. Sustain. Energy Rev.* **2015**, *44*, 847–866. [CrossRef]
16. Madanayake, B.N.; Gan, S.Y.; Eastwick, C.; Ng, H.K. Biomass as an energy source in coal co-firing and its feasibility enhancement via pre-treatment techniques. *Fuel Process. Technol.* **2017**, *159*, 287–305. [CrossRef]

17. Koppejan, J.; Cremers, M.; Middelkamp, J.; Witkamp, J.; Sokhansanj, S.; Melin, S.; Madrali, S. *Status Overview of Torrefaction Technologies: A Review of the Commercialisation Status of Biomass Torrefaction*; International Energy Agency Bioenergy: Paris, France, 2015; Available online: http://www.ieabcc.nl/publications/IEA_Bioenergy_T32_Torrefaction_update_2015b.pdf (accessed on 9 October 2017).

18. Bergman, P.C.A.; Boersma, A.R.; Zwart, R.W.R.; Kiel, J.H.A. *Torrefation for Biomass Co-Fring in Existing Coal-Fired Power Stations "Biocoal"*; ECN-C-05-013; Energy Research Centre of the Netherlands: Sint Maartensvlotbrug, The Netherlands, 2005.

19. Du, S.-W.; Chen, W.-H.; Lucas, J.A. Pretreatment of biomass by torrefaction and carbonization for coal blend used in pulverized coal injection. *Bioresour. Technol.* **2014**, *161*, 333–339. [CrossRef] [PubMed]

20. Phanphanich, M.; Mani, S. Impact of torrefaction on the grindability and fuel characteristics of forest biomass. *Bioresour. Technol.* **2011**, *102*, 1246–1253. [CrossRef] [PubMed]

21. Repellin, V.; Govin, A.; Rolland, M.; Guyonnet, R. Energy requirement for fine grinding of torrefied wood. *Biomass Bioenergy* **2010**, *34*, 923–930. [CrossRef]

22. Arias, B.; Pevida, C.; Fermoso, J.; Plaza, M.G.; Rubiera, F.; Pis, J.J. Influence of torrefaction on the grindability and reactivity of woody biomass. *Fuel Process. Technol.* **2008**, *89*, 169–175. [CrossRef]

23. Prins, M.J.; Ptasinski, K.J.; Janssen, F. More efficient biomass gasification via torrefaction. *Energy* **2006**, *31*, 3458–3470. [CrossRef]

24. Liu, Z.G.; Han, G.H. Production of solid fuel biochar from waste biomass by low temperature pyrolysis. *Fuel* **2015**, *158*, 159–165. [CrossRef]

25. Davidsson, K.O.; Pettersson, J.B.C. Birch wood particle shrinkage during rapid pyrolysis. *Fuel* **2002**, *81*, 263–270. [CrossRef]

26. Kumar, R.R.; Kolar, A.K.; Leckner, B. Shrinkage characteristics of casuarina wood during devolatilization in a fluidized bed combustor. *Biomass Bioenergy* **2006**, *30*, 153–165. [CrossRef]

27. Kwiatkowski, K.; Bajer, K.; Celinska, A.; Dudynski, M.; Korotko, J.; Sosnowska, M. Pyrolysis and gasification of a thermally thick wood particle—Effect of fragmentation. *Fuel* **2014**, *132*, 125–134. [CrossRef]

28. Chen, W.H.; Cheng, W.Y.; Lu, K.M.; Huang, Y.P. An evaluation on improvement of pulverized biomass property for solid fuel through torrefaction. *Appl. Energy* **2011**, *88*, 3636–3644. [CrossRef]

29. Mafu, L.D.; Neomagus, H.; Everson, R.C.; Carrier, M.; Strydom, C.A.; Bunt, J.R. Structural and chemical modifications of typical south african biomasses during torrefaction. *Bioresour. Technol.* **2016**, *202*, 192–197. [CrossRef] [PubMed]

30. Wang, P.; Shuster, E.; Matuszewski, M.; Tarka, T.; VanEssendelft, D.; Berry, D. Selection of biomass type for co-gasification studies. In Proceedings of the 35th International Technical Conference on Clean Coal & Fuel Systems, Clearwater, FL, USA, 6–10 June 2010.

31. Rowell, R.M.; Pettersen, R.; Han, J.S.; Rowel, J.S.; Tshabalala, M.A. Cell wall chemisty. In *Handbook of Wood Chemistry and Wood Composites*; Rowell, R.M., Ed.; Tayer & Francis: New York, NY, USA, 2005.

32. Winandy, J.E.; Rowell, R.M. Chemistry of wood strength. In *Handbook of Wood Chemistry and Wood Composites*; Rowell, R.M., Ed.; Tayor & Francis: New Yok, NY, USA, 2005.

33. Hanaoka, T.; Inoue, S.; Uno, S.; Ogi, T.; Minowa, T. Effect of woody biomass components on air-steam gasification. *Biomass Bioenergy* **2005**, *28*, 69–76. [CrossRef]

34. Serapiglia, M.J.; Cameron, K.D.; Stipanovic, A.J.; Smart, L.B. High-resolution thermogravimetric analysis for rapid characterization of biomass composition and selection of shrub willow varieties. *Appl. Biochem. Biotechnol.* **2008**, *145*, 3–11. [CrossRef] [PubMed]

35. Chen, W.H.; Kuo, P.C. A study on torrefaction of various biomass materials and its impact on lignocellulosic structure simulated by a thermogravimetry. *Energy* **2010**, *35*, 2580–2586. [CrossRef]

36. Bahng, M.-K.; Mukarakate, C.; Robichaud, D.J.; Nimlos, M.R. Current technologies for analysis of biomass thermochemical processing: A review. *Anal. Chim. Acta* **2009**, *651*, 117–138. [CrossRef] [PubMed]

37. Fisher, E.M.; Dupont, C.; Darvell, L.I.; Commandre, J.M.; Saddawi, A.; Jones, J.M.; Grateau, M.; Nocquet, T.; Salvador, S. Combustion and gasification characteristics of chars from raw and torrefied biomass. *Bioresour. Technol.* **2012**, *119*, 157–165. [CrossRef] [PubMed]

38. Demirbas, A. Biorefineries: Current activities and future developments. *Energy Convers. Manag.* **2009**, *50*, 2782–2801. [CrossRef]

39. Friedman, J. *Wood Identification by Microscopic Examination: A Guide for the Archaeologist on the Northwest Coast of North America (Heritage Record)*; British Columbia Provincial Museum: Victoria, BC, Canada, 1978.

40. Erol, M.; Haykiri-Acma, H.; Kucukbayrak, S. Calorific value estimation of biomass from their proximate analyses data. *Renew. Energy* **2010**, *35*, 170–173. [CrossRef]

41. Demirbas, A. Relationships between lignin contents and heating values of biomass. *Energy Convers. Manag.* **2001**, *42*, 183–188. [CrossRef]

42. Parikh, J.; Channiwala, S.A.; Ghosal, G.K. A correlation for calculating hhv from proximate analysis of solid fuels. *Fuel* **2005**, *84*, 487–494. [CrossRef]

43. Yang, H.; Yan, R.; Chen, H.; Lee, D.H.; Zheng, C. Characteristics of hemicellulose, cellulose and lignin pyrolysis. *Fuel* **2007**, *86*, 1781–1788. [CrossRef]

44. Wiedenhoeft, A.C.; Miller, R.B. Structure and function of wood. In *Handbook of Wood Chemistry and Wood Composites*; MRowell, R.M., Ed.; Taylor & Francis: New York, NY, USA, 2005; pp. 9–33.

45. Byrne, C.E.; Nagle, D.C. Carbonized wood monoliths—Characterization. *Carbon* **1997**, *35*, 267–273. [CrossRef]

46. Yan, W.; Acharjee, T.C.; Coronella, C.J.; Vasquez, V.R. Thermal pretreatment of lignocellulosic biomass. *Environ. Prog. Sustain. Energy* **2009**, *28*, 435–440. [CrossRef]

47. Byrne, C.E.; Nagle, D.C. Carbonization of wood for advanced materials applications. *Carbon* **1997**, *35*, 259–266. [CrossRef]

48. Vassilev, S.V.; Vassileva, C.G.; Vassilev, V.S. Advantages and disadvantages of composition and properties of biomass in comparison with coal: An overview. *Fuel* **2015**, *158*, 330–350. [CrossRef]

49. Haas, T.J.; Nimlos, M.R.; Donohoe, B.S. Real-time and post-reaction microscopic structural analysis of biomass undergoing pyrolysis. *Energy Fuels* **2009**, *23*, 3810–3817. [CrossRef]

Article

Torrefied Biomass Pellets—Comparing Grindability in Different Laboratory Mills

Jan Hari Arti Khalsa [1,*], Diana Leistner [1], Nadja Weller [1], Leilani I. Darvell [2] and Ben Dooley [2]

[1] Deutsches Biomasseforschungszentrum gemeinnützige GmbH, Leipzig 04347, Germany; diana.leistner@web.de (D.L.); nadja.weller@dbfz.de (N.W.)
[2] School of Chemical and Process Engineering, University of Leeds, Leeds LS2 9JT, UK; l.i.darvell@leeds.ac.uk (L.I.D.); pm07bd@leeds.ac.uk (B.D.)
* Correspondence: jan.khalsa@dbfz.de; Tel.: +49-341-2434-396

Academic Editor: Jaya Shankar Tumuluru
Received: 17 August 2016; Accepted: 26 September 2016; Published: 4 October 2016

Abstract: The firing and co-firing of biomass in pulverized coal fired power plants around the world is expected to increase in the coming years. Torrefaction may prove to be a suitable way of upgrading biomass for such an application. For transport and storage purposes, the torrefied biomass will tend to be in pellet form. Whilst standard methods for the assessment of the milling characteristics of coal exist, this is not the case for torrefied materials—whether in pellet form or not. The grindability of the fuel directly impacts the overall efficiency of the combustion process and as such it is an important parameter. In the present study, the grindability of different torrefied biomass pellets was tested in three different laboratory mill types; cutting mill (CM), hammer mill (HM) and impact mill (IM). The specific grinding energy (SGE) required for a defined mass throughput of pellets in each mill was measured and results were compared to other pellet characterization methods (e.g., durability, and hardness) as well as the modified Hardgrove Index. Seven different torrefied biomass pellets including willow, pine, beech, poplar, spruce, forest residue and straw were used as feedstock. On average, the particle-size distribution width (across all feedstock) was narrowest for the IM (0.41 mm), followed by the HM (0.51 mm) and widest for the CM (0.62 mm). Regarding the SGE, the IM consumed on average 8.23 Wh/kg while CM and HM consumed 5.15 and 5.24 Wh/kg, respectively. From the three mills compared in this study, the IM seems better fit for being used in a standardized method that could be developed in the future, e.g., as an ISO standard.

Keywords: grindability; torrefied biomass; pellet; energy consumption; co-firing

---

1. Introduction

Over the past few years there has been a significant effort worldwide to increase the utilization of renewable energy sources for electricity and heat production. The European Union Renewables Directive, for example, set a target of 20% of the energy consumption of the European Union to be from renewable sources by 2020 [1] and current decarburization scenarios point at 30% by 2030 and at least 55% by 2050 [2]. In this context, the firing of biomass and the co-firing of biomass with coal in power plants originally designed for coal play a significant role. This approach to the generation of renewable energy has been demonstrated in more than 200 power plants over the recent years taking advantage of the infrastructure already in place within the electricity supply industry, and the associated low capital investment requirements [3].

Besides the expected reduction in both NO_x and SO_x levels, there are still some difficulties in using raw biomass for co-firing with coal [4]. Amongst the biggest challenges are the diverse chemical and physical properties of biomass that need to be addressed. The ash content as well as volatiles and oxygen content can be higher in biomass than in coal resulting in lower gross calorific value of the fuel [5]. Additionally, problems may occur if feeding and milling is not adapted. The particle size distribution of ground biomass is different from coal and particles may be considerably larger leading to lower burn-out efficiencies [6]. Overall, co-firing with regular biomass may lead to instability of the combustion process making biomass with coal-like properties the most desirable co-firing fuel.

One intensively researched approach for upgrading regular biomass for co-firing with coal is the thermo-chemical pre-treatment best known as torrefaction. At temperatures in the range of 250–300 °C, and in the absence of air, the biomass undergoes a mild form of pyrolysis and is converted into a solid fuel with chemical and mechanical properties similar to coal [7]. This process has been studied on a large variety of feedstock applying different torrefaction settings (e.g., temperature, and time) mainly on lab-scale [8] and most recently also in pilot scale plants [9]. Several technological innovations were made over the past years to bring the process to commercial scale [10] and techno-economic assessments assuming European and Northern American conditions were conducted [11,12]. In both cases, the now more homogenous, brittle and hydrophobic material is compacted to pellets and can be transported, stored for prolonged periods and delivered as a boiler fuel.

Some experts came to the conclusion that the percentage of biomass for co-firing in power plants can be as high as 40% when the torrefied biomass is used [13] while others showed that, according to CFD simulations, coal boilers can handle operation with up to 100% of torrefied biomass without an obvious decrease of energy efficiency and fluctuation of the boiler load [14]. The direct application of torrefied biomass pellets in small-scale pellet boilers may not be advisable at this point unless boiler technology is adapted [15].

For the utilization of torrefied biomass in pulverized coal fired plants it is necessary to assess its fuel properties (chemical and mechanical) beforehand. For most of the parameters standardized analysis methods are available. However, the difference in characteristics between torrefied biomass and coal requires some adaptation of existing methods or even new methods to fully describe their fuels properties, which would then allow comparison with those of coals. A power plant operator will need to, e.g., adjust the mills for achieving the right particle size distribution for optimal plant operation. It is assumed that particle size distribution of the milled material may effect combustion efficiency, the amount of unburned carbon in the ash and the stability of combustion [16]. The Hardgrove Index (HGI), developed in the 1930 by Hardgrove [17], was designed to compare grindability of brittle materials such as black coal. It is commonly used as a simple empirical laboratory procedure for the characterization of their milling behaviour in large coal mills to produce a pulverized fuel with the appropriate fineness. The HGI method described in ISO 5074 [18], however, is not suitable for the characterization of more fibrous material like torrefied biomass and does not consider the actual milling of pellets.

Several studies have determined the grindability of torrefied biomass by applying one of the following methods: the modified Hardgrove Index [16], the Grindability Criterion (GC) [19], the Hybrid Work Index (HWI) [20] and the Impact Grindability Index (PMI) [21]. The HGI was developed for the testing of black coals, which are commonly milled in ball and roller mills. The PMI was developed for low rank coals, which are commonly processed in beater or fan mills. The modified HGI, the HWI and the GC were specifically designed to characterize the grindability of torrefied biomass. A number of boiler makers and mill suppliers also have scaled down coal mills that are employed for test purposes.

All of the current laboratory grindability tests require the preparation of a pre-crushed feed material of a specific particle size. In all cases, the test results provide an assessment of the relative grinding behaviour of the test materials. It may, however, be difficult to extrapolate the data from a laboratory test to provide quantitative information on the behaviour of torrefied biomass pellets in an industrial scale mill.

The aim of the work presented here, which was part of the SECTOR-project (funded under the EU FP7 programme) [9], is to test a grindability method (GM) for torrefied biomass pellets, which is simple, fast and reliable and based on the grinding of pellets rather than pre-crushed material. The method will result in a single parameter: the specific grinding energy (SGE), i.e., the energy consumed when a defined pellet mass (2.5 kg), at a defined mass flow is ground to a particle top size of 1 mm. This parameter, or the basic principles behind it, may eventually become useful to pellet producers and plant operators if grindability is introduced to a standardized product specification of thermally treated biomass, which is currently being drafted in the relevant ISO TC238 working group (as ISO/PRF TS 17225-8 [22]).

The results of the GM, when done with different mill types (hammer, cutting and impact mill), will be compared and their relation to other pellet stability measures such as the modified HGI, mechanical durability and a hardness test will be discussed. Finally, advantages and disadvantages of the different mill types as proper equipment for a standardized method are discussed.

2. Materials and Methods

2.1. Torrefied Biomass Pellets

Five different types of wood, one batch of straw and one batch of forest residue were torrefied at different temperatures and retention time. The raw materials were torrefied at different pilot plants (capacity <200 kg/h) with slightly different technology: willow and forest residue material was torrefied in northern Sweden (conveying screw reactor); the pine, poplar, beech and straw material in Spain (rotating shaft reactor) and the spruce material in the Netherlands (moving bed reactor). All torrefied material was compressed into pellets on site with the available pelletizing technology resulting in two different pellet diameters (6 and 8 mm).

Most relevant fuel characteristics (chemical parameters), pellet characteristics (physical-mechanical parameters) and torrefaction characteristics (temperature and time) were analysed (via one representative sample of each pellet type) according to standard methods (if available) and are presented and discussed in Section 3.1. The degree of torrefaction (TF_{dgr}), for which no standardized method exists yet, was defined here as the reduction in volatile matter content (VM) caused by the torrefaction process, based on dry matter (DM) values, according to Equation (1):

$$TF_{dgr} = 100 - \left(\frac{VM_{before} \ (\%DM)}{VM_{after} \ (\%DM)} \times 100 \right) \tag{1}$$

The pellet length is an average value of 20 randomly sampled pellets from the bulk as was required for the method described in Section 2.2.

2.2. Hardness Test

The hardness (also compressive resistance) is defined as the resistance to fracture of the pellets and is supposed to help characterize the pellets regarding their stability during e.g., storage or handling [23]. The pellet hardness values were measured using a Pellet Hardness Tester with display and PC interface (Amandus Kahl GmbH, Reinbek, Germany) which was designed for single pellets with $\varnothing$ = 3–14 mm. In this test, the pellet rests in a flat position on an anvil and is stressed by a plunger with a flat tip with a diameter of 5 mm (see Figure S1).

In order to achieve a representative value, the following procedure was applied: At first 20 pellets were randomly selected and the average length (L_{avg}) was determined (data shown in Section 3.1). Based on the L_{avg} another 20 pellets in the range of L_{avg} ± 2 mm were selected. These 20 pellets were the actual pellets being individually placed on the anvil of the Hardness Tester. Each pellet was exposed to the plunger pressing into the pellets with a speed of 5 mm/min and the increase in force (given in N) was recorded (with 100 values per second). The actual hardness value is the maximum value of N, which occurs right before the pellet fractures and a sudden decline in N is observed. From each set of 20 hardness values the min and max value was removed to smooth the distribution and an adjusted mean value was calculated (see results in Section 3.2).

2.3. Modified Hardgrove Index

The standard procedure is described in ISO 5074 [18]. A specialized ball mill apparatus (BM) is calibrated using reference coals with known HGI values. A defined mass of coal (50 g) with a specific particle size range (1.18 and 0.6 mm) is then ground and sieved, and the mass fraction of the particles passing through the 75-μm sieve is determined. From this value the HGI is calculated. The lower the HGI value the harder and less grindable the material is.

Bridgeman et al. [16] recognized that this method needs to be adapted for biomass because the volumes of biomass and coals differ significantly receiving the same grinding energy in the mill [24,25] and results will thus not be comparable. Instead of using a defined mass the modified HGI (mHGI) requires a defined volume (50 cm^3) of the substrate to be ground. Thus, the torrefied biomass pellets were pre-crushed and sieved according to ISO 5074 and treated following the method laid out by Bridgeman et al. [16] with an initial calibration of the mill using four standard coals with known HGI values. Mean values of the duplicates were calculated and are used for further discussion in this study.

2.4. Specific Grinding Energy

2.4.1. Experimental Setup

The experimental set-up to estimate the SGE consisted of the following equipment: (i) mill; (ii) feeding unit; and (iii) power measuring device.

Three common laboratory mill types for grinding of biomass where chosen each imposing a different kind of stress on the material (see pictures of the mill interior in Figure S2). In the cutting mill (CM: Pulverisette 19, FRITSCH GmbH, Idar-Oberstein, Germany), the material is predominantly cut by the blades, while, in the hammer mill (HM: CHM 230, NETZSCH GmbH, Selb, Germany), the flexible hammer elements crush the material. The impact mill (IM: Rekord A, Jehmlich GmbH, Nossen, Germany) has neither cutting elements nor flexible elements, and the grinding is caused by the specific geometry of a disc (grinding ring) rotating in the chamber. For each mill an outlet sieve with mesh size Ø = 1 mm is used. For tests with HM and IM a round-hole sieve is used and for the CM a sieve with trapezoidal meshes is used. All mills are constructed for grinding soft to medium-hard or fibrous biogenic feedstock. The technical specifications are summarized in Table 1.

Table 1. Technical specification of the mills (as given by the manufacturer) and realized feeding rates of the metering unit given the specific fuel type.

Technical Specs		Impact Mill (IM)	Cutting Mill (CM)	Hammer Mill (HM)
Throughput	kg/h	200	60	50–100
Rotor speed	rpm	12000	3000	3600
Engine power	kW	7.5	1.5	5.5
Feeder Setting				
mean	Hz	16.7	16.7	16.2
min	Hz	10.0	10.0	13.4
max	Hz	21.2	21.2	18.1

In order to feed the pellets at a uniform mass flow into the mill a screw feeder (GLD 87, Gericke AG, Regensdorf, Switzerland) was used. On average the feeder was operated with 16.2 (HM) and 16.7 Hz (IM, and CM) realizing a comparable pellet mass flow between the different sets of experiments. However, due to the difference in bulk density, length and diameter individual feeding rates were estimated for the different pellet types, which could vary between 10 and 21.2 Hz (see Table 1).

A clamp meter (345 PQ, Fluke Corporation, Everett, WA, USA) was used as the power measuring device and set to record one value per second.

2.4.2. Procedure

Prior to the actual measurements the following preparatory measures were taken: the fines of the pellets were screened with a 3.15 mm round-hole sieve (according to EN 15149-1); a 2.5 kg sample (with moisture as received) was prepared for the measurement; the feeder was adjusted to the previously estimated setting; the power measuring device was installed directly at the power supply line of the mill; and the mill was equipped with 1 mm mesh.

A full recording cycle (see Figure S3) was comprised of: a mill idling period of at least 2 min, during which the idle power consumption of the mill running empty was recorded; the actual milling period; and a second idle period of at least 2 min after no more material fell through the sieve.

The ground material was weighed to check on whether grinding was complete. The whole procedure was repeated at least once for each pellet type.

2.4.3. Calculating the Specific Grinding Energy

The determination of SGE involves the measurement of the idling power before grinding (P_{I1}), the total active power during milling process (P_{total}), the idling power after the grinding (P_{I2}), the actual duration of the grinding (t_G) and the mass of material that was ground (m_G). Preliminary tests showed that differences between P_{I1} and P_{I2} might occur; hence, both values are necessary. In order to make the measurement comparable across all samples it was decided to define t_G as the time when P_{total} is at a stable plateau. Even though minimal grinding takes place before and after that (grey areas in Figure S3), it was found to be the simplest and most representative way of reading the P_{total} from the power curve.

As P_{total} is the sum of the power applied for grinding and the idle power the actual power needed just for the grinding (P_G) is calculated by subtracting the averaged idle powers of P_{I1} and P_{I2} (P_I) from P_{total}.

The average idle power P_I (in W):

$$P_I = \frac{P_{I1} + P_{I2}}{2} \tag{2}$$

The power needed for grinding the pellets P_G (in W):

$$P_G = P_{total} - P_I \tag{3}$$

By multiplying P_G with t_G (in h) the grinding energy (E_G, in Wh) is calculated:

$$E_G = P_G \cdot t_G \tag{4}$$

Finally, SGE (in Wh/kg) is defined as:

$$SGE = \frac{E_G}{m_G} \tag{5}$$

2.5. Particle Size Distributions

For the comparison of the three mill types and their grinding performance, the particle size distributions (PSD) were obtained after sieving the ground material according to DIN EN 15149-2 with sieve mesh size of 3.15, 2.00, 1.00, 0.71, 0.50, 0.40, 0.25, 0.16, 0.10 and 0.063 mm. The data points used for the distribution curves refer to the cumulative mass which was observed under the respective mesh size, e.g., the mass obtained at a particle size value of 0.40 mm are all the particles that fell through the sieves with mesh size 0.40 mm and under.

A representative particle size value for which 10%, 50% and 90% of the mass are accumulated (x10, x50, x90) can be calculating from: (i) the upper (X_u) and the lower (X_l) size class value (in mm) between which the respective particle size value is to be expected, (ii) the mass percentage (%) under each of the two size class (Q_u and Q_l) and (iii) the ratio (R) between ($Q_u - Q_l$) and ($X_u - X_l$). The following equations were used:

$$R = \frac{Q_u - Q_l}{X_u - X_l} \tag{6}$$

$$x90 = X_l + \frac{90 - Q_u}{R} \tag{7a}$$

$$x50 = X_l + \frac{50 - Q_u}{R} \tag{7b}$$

$$x10 = X_l + \frac{10 - Q_u}{R} \tag{7c}$$

The difference between x90 and x10 is defined here as the width of the PSD curve and used in Section 3.4 to discuss the characteristics of the distribution curves. The full dataset, however, can be found in Table S1 of the Supplementary Materials.

3. Results and Discussion

3.1. General Fuel Charactersitic of the Torrefied Biomass Pellets

The seven different biomass materials all underwent slightly different torrefaction conditions with a TF_{temp} range between 260 °C and 308 °C while the residence time of the biomass in the reactor varied between 9 and 30 min (see Table 2). In order to compare the relative change of the material caused by torrefaction, the change in volatile content was chosen as TF_{dgr}, which, in this case, ranged between 6 and 9% DM. Regarding the HHV, which is increased during torrefaction, values found by other torrefaction studies were in some cases similar to what was obtained in this study and in other cases not. Gucho et al. [26] for example found torrefied beech wood both at 260 °C and at 280 °C for 30 min and the mean HHV between the two is the same as the value obtained here for 30 min at 270 °C. Phanphanich and Mani [27] on the other hand torrefied pine chips at 300 °C for 30 min and obtained a HHV of 25.4 MJ/kg, much higher than the 21.9 MJ/kg observed here. However, both studies torrefied on lab scale in batch reactors and it is known from studies on biomass pyrolysis that particles size and sample size, let alone the reactor type, will have an impact on the product [28]. Reports on increasing HHV with increasing intensity of torrefaction when using pilot scale plants are available and support the findings here [29,30].

Table 2. Torrefaction conditions, fuel characteristics after torrefaction and pellet characteristics of the seven torrefied raw-materials.

Parameter	Unit	Method (ISO)	Forest Residues	Willow	Pine	Poplar	Spruce	Beech	Straw
Torrefaction conditions			-	-	-	-	-	-	-
TF_{temp}	°C	-	308	308	300	280	260	270	270
TF_{time}	min	-	9	9	30	30	- *	30	30
TF_{dgr}	% DM	-	9	6	8	8	6	6	7
Fuel characteristics			-	-	-	-	-	-	-
HHV	MJ/kg DM	18125	23.3	21.2	21.9	21.4	21.5	20.2	20.4
Ash	% DM	18122	3.1	2.2	0.6	1.5	0.3	1.4	4.5
Volatiles	% DM	18123	67.9	74.1	77.1	76.7	78.6	79.3	72.6
C	% DM	16948	53.6	51.7	52.6	51.1	50.7	49.4	49.0
H	% DM	16948	6.5	6.6	6.4	6.4	5.9	6.3	6.8
N	% DM	16948	0.6	0.3	0.1	0.1	0.1	0.2	0.6
S	% DM	16994	0.036	0.021	0.009	0.011	0.004	0.011	0.075
Cl	% DM	16994	0.005	0.001	0.001	0.005	0.002	0.003	0.050
Pellet characteristics			-	-	-	-	-	-	-
Water	% FM	18134-1	5.1	10.5	6.7	6.9	2.2	7.3	10.4
Bulk density	kg/m³	17828	679	592	631	610	694	634	727
Mech. durability	%	17831-1	89.3	87.7	90.3	94.5	97.6	95.7	97.9
Ø	mm	-	8	8	6	6	8	6	6
Average length	mm	-	10.2	8.0	13.0	17.0	10.0	21.0	22.5

* not recorded.

Specific requirements for chemical and physical-mechanical characteristic of the pellets made from thermally treated are currently finalized in a technical specification (TS) on international level as ISO/PRF TS 17225-8 [22]. Once finalized, ISO/PRF TS 17225-8 will have an outline much like the ISO 17225-2 [31] for pellets made from wood or ISO 17225-6 [32] for pellets made from non-woody biomass. However, as the market of thermally treated biomass is still developing a TS will serve as a placeholder until a full standard can be agreed upon. The main categorization in this TS is between woody biomass of energy content (HHV) below or above 21 MJ/kg DM. Again, there is a differentiation in woody and non-woody biomass and three respective quality classes, however, for non-woody biomass, the 21 MJ/kg DM threshold does not exist. In order to produce a fuel that is above or below 21 MJ/kg DM, the torrefaction plant operators will have to decide how they run their reactor accordingly. In case of this study, which took place before the conception of this TS, almost all woody fuels are above 21 MJ/kg DM (except for the beech pellets) which is why the requirements for bulk density, durability, moisture and ash will be discussed based on the thresholds that apply to that category. None of the here investigated woody biomass pellets could meet the criteria for the highest quality class. Even the second highest class, which requires bulk density to be >650 kg/m^3, durability to be >96%, moisture to be <8% water content, and ash to be <3%, could only be met by spruce pellets (see Table 2). However, bulk density, durability, and moisture of the pellets are all parameters that are a result of pelletizing process while the HHV is a result of the torrefaction process. Creating standardized high quality fuel pellet will only be a problem when ash content of the input material is already high as it will only be increased during torrefaction. The straw pellets produced here were able to meet many of the requirements for the highest quality class of pellets made from thermally treated non-woody biomass such as bulk density (>600 kg/m^3), durability (>97.5%, see optimization of durability in [9]) and ash content (<5%). However, the S content was posing a problem as it was above the threshold of 0.05% for the highest quality class. In addition, the moisture was slightly above the threshold of 10%, which is no major challenge and can be adjusted either during pelletizing or through subsequent drying. The S content, however, is much like the ash content a parameter that needs to be controlled in the raw material as the torrefaction will not help reduce it. In general, a non-woody biomass pellet to fall into the thermally treated biomass standard not only needs to be thermally treated but should also have a minimum HHV of 18 MJ/kg DM for the highest quality class.

3.2. Hardness and Modefied Hardgrove Index of the Torrefied Biomass Pellets

Beyond the standard characteristics, which are prescribed in the respective ISO standards, the quality of the pellets may also be described by other relevant properties e.g., resistance to fracture or grindability. In this study, hardness was chosen as a parameter that may bridge the gap between the information obtained from the mechanical durability test (ISO 17831-1) to the actual grindability using a laboratory mill. In addition, the hardness test may also help understand the unintended grinding which happens when pellets are being fed, e.g., using a screw feeder. This, however, was not focus of this study.

The average hardness of the different pellets types was very different with 96 N for forest residue pellets and 496 N for spruce pellets (Table 3). This variation is partly reflected by the differences in torrefaction conditions with forest residue being exposed to the highest and spruce being exposed to the lowest TF$_{temp}$. This may be explained by the removal of hydrogen bonding sites, depolymerisation and the destruction of the fibrous structure (less interlocking bonds) with increased degree of torrefaction. The high standard deviation, which is observed when recording the pellet hardness across 20 pellets, as proposed here, may reflect the inhomogeneity of the torrefied material leading to different structural bonds in the pellets. In order to produce torrefied pellets with high durability and a respective hardness, application of steam and water prior to densification may be crucial to help plasticize the material and reduce the softening temperature of the remaining lignin [9].

In literature only one study could be found reporting the hardness of torrefied biomass pellets measured via the Kahl Hardness Tester [30], while Li et al. [33] compared the Meyer hardness of torrefied wood pellets which in addition to the force also includes the indentation depth and the surface area the force is applied to. Li et al. [33] found that the higher the severity of the torrefaction the lower the hardness of the pellets was. This is in line with the findings of this study.

Table 3. Mean values of the modified Hardgrove index and hardness of the torrefied biomass pellets as well as further statistical information on how the mean values were obtained.

Parameter		Biomass Type						
		Forest Residues	Willow	Pine	Poplar	Spruce	Beech	Straw
mHGI		-	-	-	-	-	-	-
Mean		55	34	36	28	25	21	55
1st run		56	34	36	27	25	22	56
2nd run		54	34	36	28	26	20	54
Hardness (N)		-	-	-	-	-	-	-
Adjusted	Mean (n = 18)	96	103	125	206	496	275	273
	sd	37	22	31	59	79	52	48
	min	52	65	94	101	329	197	211
	max	189	140	212	299	642	351	371
Original Mean (n = 20)		98	105	128	205	486	272	273

Regarding a comparison with standardized Hardgrove coals by applying the modified HGI method, forest residue and straw had the highest value (HGI = 55) which reflects a low energy input, better grindability and potentially a higher throughput while spruce and beech had the lowest HGI values (HGI = 25 and 21 respectively) being not so favourable for grinding. Similar to the method used here with 50 cm^3 samples, Shang et al. [34] investigated the HGI of torrefied wheat straw at different TF$_{temp}$ with a residence time much higher (2 h) than any of the TF$_{time}$ used here. The HGI of the pellets made from straw torrefied at 270 °C for 30 min analysed here (HGI = 55) is therefore slightly below the value observed by Shang et al. [34] for straw torrefied at 250 °C for 2 h (HGI ≈ 60). Bridgeman et al. [16] measured the modified HGI of torrefied willow and found values between 24 and 51 mainly depending on the difference in TF$_{time}$ (10 vs. 60 min) while the TF$_{temp}$ was in both cases at 290 °C. The pellets from torrefied willow analysed here had a HGI of 34 which is between the two HGIs measured by Bridgeman et al. [16] probably due to a higher TF$_{temp}$ (308 °C) at similar TF$_{time}$ (9 min). Interestingly, the results obtained by Bridgeman et al. [16] and Shang et al. [34] compare well to the results obtained here even though the torrefaction in these two studies was done in lab-scale reactors as compared to the pilot-scale torrefaction which might affect homogeneity of the product.

No evident correlation could be observed between mHGI and the hardness. This is to be expected as the hardness test gives an information on the resistance of the pellet to fracture while the mHGI is measured on the pre-ground pellet (<1.18 mm) and therefore reflects the characteristics of those particles. The hardness test is more similar to the durability test which is supported by a correlation between the two parameters of R^2 = 0.72. Durability and HGI values are also not correlated.

3.3. Performance of the Mills and Specific Grinding Energy

The range in time required to grind 2.5 kg of pellets using a 1 mm sieve varied between the different mills (Figure 1a). The smallest variation was observed for pellets ground with the IM (2.3–2.7 min) while for the CM the range was between 2.1 and 4.4 min. The type of pellet requiring the longest (t$_{max}$) or shortest time (t$_{min}$) to be ground was never the same between the mills (IM: t$_{max}$ = poplar, t$_{min}$ = beech; CM: t$_{max}$ = straw, t$_{min}$ = willow; HM: t$_{max}$ = forest residue, t$_{min}$ = pine) highlighting that different grinding forces are at work. For IM and HM, the grinding time was independent of mHGI, hardness or durability of the pellets (all R^2 < 0.01). However, the grinding time in the CM was showing signs of dependency on hardness (R^2 = 0.50) and a strong correlation (R^2 = 0.79) with durability. The fact that IM takes the least time for crushing the pellets below 1 mm is largely due to its capacity. As by construction the impact mill is designed for the highest throughput and operating at higher rotor speed than the other two mills. Given the procedure chosen in this study (2.5 kg input material) the IM was not operating at full capacity.

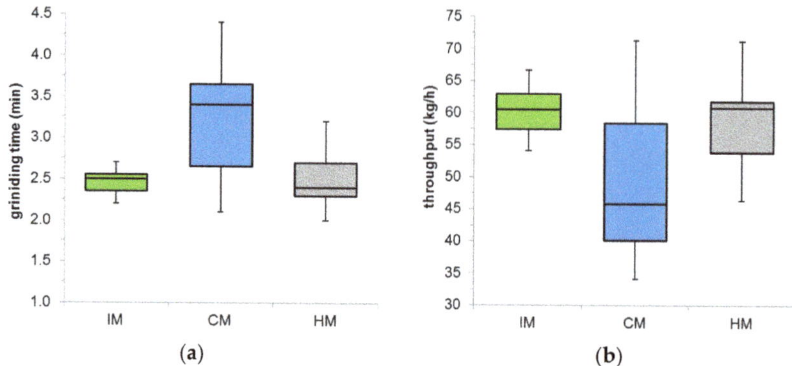

Figure 1. Boxplot for each mill type showing: (**a**) The pellet specific grinding time (t_G); (**b**) The pellets specific throughput. Boxplot shows min and max values (error bars) as well as the 1st and 3rd quartile and the median.

Regarding the throughput of the mills (Figure 1b), which by definition is dependent on the grinding time, smallest variation was observed for IM and HM while CM had a wide range (34.1–71.3 kg/h). Since low grinding time leads to a high throughput, the pattern of pellet specific throughput per mill is the reverse pattern of pellet specific grinding time mentioned above.

How the different torrefied biomass pellets resulted in different SGEs can be seen in Figure 2. The mean SGE observed for CM and HM were about the same at 5.15–5.24 Wh/kg while the mean SGE of the IM was substantially higher with 8.23 Wh/kg. Again, this reflects the similar technical specifications of both CM and HM in terms of throughput and rotor speed, which is much different from the IM. More interesting, however, is the way the different mills performed with the respective pellets. The commonality between all three mills was that forest residue pellets required the least and straw the most energy for grinding. In addition, all three mills were able to grind forest residue, willow and pine well below the average SGE consumption while in addition to straw poplar and spruce and beech (except for the CM) required above average SGE. This pattern is in line with several physical-mechanical characteristics of the pellets. Forest residue, willow and pine have much lower hardness (96–125 N) and durability (87.7%–90.3%) and higher TF_{TEMP} (300–308 °C) than the other pellet types (hardness: 206–496 N; durability: 94.5%–97.9%; TF_{TEMP}: 260–280 °C).

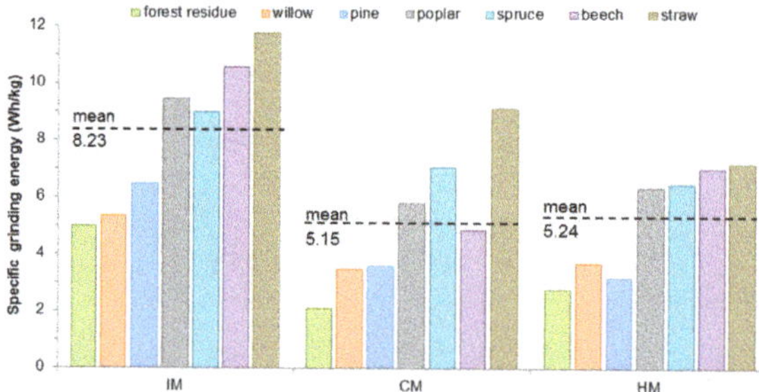

Figure 2. Specific grinding energy obtained by applying the evolved grindability method to the respective torrefied biomass pellets using three different mill types: impact mill (IM), cutting mill (CM) and hammer mill (HM).

Current literature does not allow a good classification of the results obtained here as in other studies other parameters were the centre of focus when reporting grinding energy. However, a closer look at other studies recording grinding energy may still give further insights into the results obtained here. Phanphanich and Mani [27], for example, showed a dramatic decrease in grinding energy between untreated wood chips (237 Wh/kg; at water content of 7%) and torrefied wood chips (as low as 23 Wh/kg) using a laboratory cutting mill (1.5 kW, mesh size 1.5 mm) and found a strong linear correlation with torrefaction temperature. Temmerman et al. found similar values for grinding energy of wood chips (162 Wh/kg, at water content of 5%–10%) as Phanphanich and Mani [27] using a laboratory hammer mill (1.1 kW, mesh size 2 mm), while regular wood pellets consumed energies much closer to the ones measured here for torrefied wood pellets (on average 11.4 Wh/kg; water content of 5%–10%). This may be explained by the fact that grinding pellets often is a disintegration of the pellets back into the particles they were made of. Hence, the difference in grinding energy between pellets from untreated material and torrefied material will be much smaller than between, e.g., wood chips and pellets. However, whether the grindability of untreated and torrefied material or non-torrefied and torrefied pellets are compared, increasing the degree of torrefaction should result in lower SGE. Shang et al. [35] reported an exponential decrease of the grinding energy for pine pellets exposed to increasingly higher torrefaction temperatures when using a coffee grinder as laboratory mill (0.55 kW; similar grinding forces as the IM). However, a comparison between the grindability of torrefied pellets and pellets from torrefied material may only partly apply (see discussion on bonding mechanisms during densification in Section 3.2).

The fact that pellets from torrefied material require a much lower grinding energy is one of the economic advantages of torrefaction as it will save energy during the final grinding for injecting the biomass powder into the boiler. However, the possible difference that may occur between different laboratory mill types, as shown here, may only partly be relevant for power plant operators since existing technology may not be replaced for new technology due to high investment costs. However, the results shown here could help identify the most appropriate mill type for a standardized methodology to assess grindability of regular and thermally-treated biomass pellets. The SGEs observed in the IM, for example, are much more distinct for the different pellet types than the values observed with the HM where poplar, spruce, beech and straw only differ minimally.

3.4. Particle Size Distributions after Grinding

In addition to the SGE, the cumulative PSD curves (Figure 3) were compared to assess the impact of the mills on the material and to check whether a particular mill type might prove more suitable for use in a standardized methodology in relation to, e.g., co-combustion requirements.

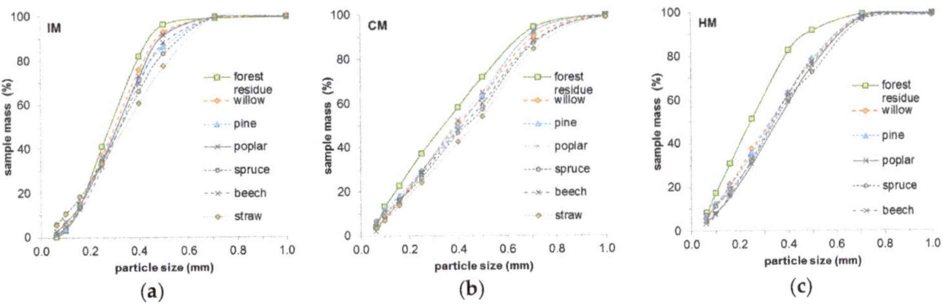

Figure 3. Cumulative particle size distribution obtained when using: (**a**) Impact mill; (**b**) Cutting mill; and (**c**) Hammer mill on each of the torrefied biomass pellets.

As can be seen in Figure 3, each mill yields a uniquely shaped distribution curve, which is similar for all materials ground with that mill. In the current literature, a comparison of PSD curves between different lab-size mills and their performance with torrefied pellets is missing. Some studies, however, may be helpful to further understand the results shown here. Phanphanich and Mani [27], for example, produced cumulative PSDs of two materials (pine chips and logging residue) with different degrees of torrefaction which were ground in a lab-size cutting mill. The resultant PSD curve showed a linear increase—much like the one in this study for the untreated materials (Figure 3b)—where it can be seen that as the severity of the torrefaction increased the incline of the curve was more pronounced at <0.25 mm before it started declining (at >0.25 mm). However, it is difficult comparing the grindability of wood chips with the grindability of pellets. Thrän et al. [9] showed that pellets from torrefied spruce were mostly disintegrated back to the particle sizes before densification. The disintegration of non-densified material is of course very different and may depend even more on the mill type. However, when looking at the three different mill types and their respective PSD curves it becomes clear that the pellets were not just disintegrated but fractioned differently. Especially the PSD curves of the IM (Figure 3a) had a unique characteristic with a slow incline up until 0.2 mm followed by a strong incline between 0.2 and 0.5 mm and phasing out thereafter. For both CM and HM the incline is steady up until 0.7 mm before the curves phase out (with the exception of forest residue in the HM). Bridgman et al. [16] showed cumulative PSD curves for two energy crops (willow and miscanthus) which where torrefied and ground in a roller-ball mill designed for the HGI test and they had very different characteristics than the curves of the three mills used here. Interestingly, in their study, all degrees of torrefaction of miscanthus resulted in curves similar to the ones from the HGI coals while the raw material yielded an entirely different curve. However, in the case of willow, only the PSD curve of the most severely torrefied sample showed the same pattern as all the standard coal curves. Despite the differences in mills, the PSD curves obtained by their study do not compare well with the result obtained here as they show the grindability of pre-ground material (as required by the HGI procedure) while here whole pellets were used. Only during the investigation of the mHGI of the seven pellet (when using a ball mill after pre-grinding the pellets) their findings (PSD curve of torrefied willow) could be confirmed (see Figure S4).

Regarding the grinding performance and suitability of the PSD for co-combustion application Thrän et al. [9] investigated the co-milling of torrefied pellets with coal in a bowl mill (400 kg/h at medium load, vertical rollers) at different mixture rates. When the share of torrefied pellets in the mixture was high (58.5%) 90% of the ground material had a particle size below 0.6 mm while at a lower share of torrefied pellets (32%) the main particle size fraction dropped below 0.2 mm. In comparison to the mills used here, only the IM can achieve similar results with most of the seven feedstocks resulting in 90% of the ground mass to be below 0.6 mm.

To assess the suitability of the mills to test grindability in a standardized way for an application in co-combustion process the distribution widths of the PSD curves were compared. The average distribution width (Figure 4) across all input materials was narrowest for the IM (0.41 mm), followed by the HM (0.51 mm) and the CM (0.62 mm). However, the IM had the largest variation in feedstock-based distribution width ranging from 0.33 mm all the way to 0.52 mm. Both HM (0.41–0.56 mm) and CM (0.57–0.69 mm) had more even distribution widths across the different feedstock. Looking back at the distribution curves again (Figure 3), it becomes obvious that the variation in the curves from the IM happened in the particle range of >0.3 mm, while the variation observed in the other two mills was more or less consistent along the particle size scale. The distribution curves of HM, on the other hand, had one strong outlier (forest residue) compared to the other mills. However, in all mills the distribution width was below average when grinding forest residue material, which is also in line with the low SGE required to grind this type of pellet in all mills. Along the same line, straw resulted in the largest distribution width in all mills (curve for straw in HM is missing from the data) as suggested by the SGE. However, a modest linear correlation of the SGEs and the respective distribution width could be observed for the CM ($R^2 = 0.65$) and the IM ($R^2 = 0.47$) while the HM had only weak correlations ($R^2 = 0.27$). Thus, for CM and IM it can be concluded that pellets from torrefied material, which generally require lower SGEs, also will be ground with narrower distribution widths. The fact that the IM not only has the narrowest distribution width but also the lowest values for x90 (0.53 mm, average across all seven pellet types; see Table S1) makes it a more ideal lab-mill for producing particles favourable in co-combustion settings.

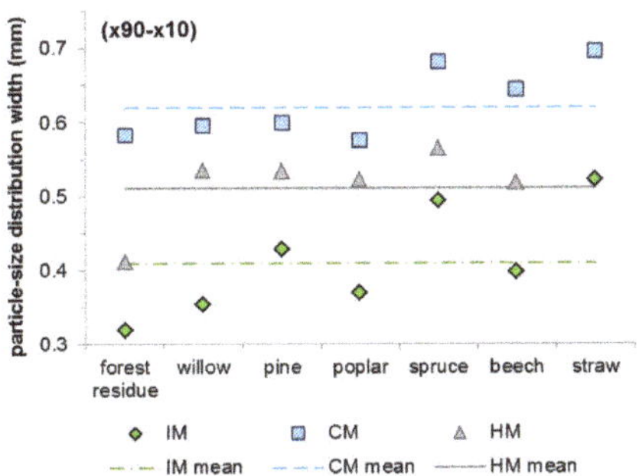

Figure 4. Width of each particle size distribution obtained from the seven torrefied biomass pellets when milled with the respective mill type: impact mill (IM), cutting mill (CM) and hammer mill (HM). The mean value across all distribution width values of each mill type is indicated.

3.5. Specific Grinding Energy versus Other Methods

To assess whether the SGE is a parameter that can stand alone or is better complemented with another parameter the SGE values were correlated to the three relevant grindability parameters analysed here. The SGE was found to be in a positive linear relation with the durability of the pellets and the pellet hardness (Table 4). However, only the correlation with the durability is of strong enough significance (R^2 = 0.86 for both IM and HM) that one could argue that SGE is partly reproducing that information. All correlations of SGE with the modified HGI were weaker ($R^2 \leq 0.77$) than the correlation between durability and SGE. The correlations between mHGI and SGE were all negative, as a lower value mHGI will imply higher energy requirement for grinding. The correlations of SGE with hardness were all of insignificant strength.

Table 4. Correlation coefficient (R^2) and equation for predicting specific grinding energy (SGE) with either modified Hardgrove Index (mHGI), hardness or durability (DU) depending on the mill type.

Mill type	SGE (Wh/kg)					
	x = mHGI	R^2	x = Hardness (N)	R^2	x = DU (%)	R^2
Cutting mill	$-0.16x + 10.13$	0.58	$0.01x + 2.46$	0.50	$0.51x - 42.58$	0.79
Impact mill	$-0.19x + 14.24$	0.70	$0.01x + 5.58$	0.40	$0.59x - 46.80$	0.86
Hammer mill	$-0.15x + 9.84$	0.77	$0.01x + 2.99$	0.54	$0.43x - 34.86$	0.86

Looking at the close correlation between SGE and durability it might be advisable to combine SGE values with a PSD parameter such as the distribution width to help detangle it from durability and give a unique insight into the co-combustion characteristics of the torrefied biomass pellets.

3.6. Adjustments of the Grindablity Method for Reliable Performance

For a standardized method, it is essential for the equipment used to perform reliably and a procedure that ensures high repeatability and reproducibility of the results obtained. For the proposed GM and the values it will yield, it is necessary to not only define the type of mill but also to adjust some parameters, e.g., the feeding rate. This will require a round robin test of the methodology across different laboratories all using the same equipment.

In terms of choosing the right mill technology for a standardized method, the transferability of the results to plant operator scenarios is the main focus. As plant operators will try to use their existing equipment (in most cases ball, roller or impact mills) and adapt it to the new biomass, a CM seems to be an unlikely choice. Another disadvantageous feature of a CM, also in terms of repeatability and reproducibility of laboratory results, are the knives, which tend to become dull and thereby bias the results (e.g., the sharpness of the knives used here could not be quantified). A HM on the other hand may pose the challenge of proper adjustment of the hammers so that results will be repeatable and reproducible, which, however, is less of a challenge than what the CM poses. IM seem to be the easiest to handle and adjust and yielded the most promising PSD curves and distinct SGE values amongst the three mills. Results from a mill like Thrän et al. [9] used in there study may be even more valuable to plant operators yet it is not applicable for a lab standard given its capacity (double that of the IM) and respective investment cost. As the IM used here is still relatively large for lab application, it might be advisable to also do an investigation on the performance of a small-scale disc-grinding system as was used by Shang et al [35].

4. Conclusions

The proposed method for assessing the specific grinding energy of torrefied biomass pellets is a promising tool that could allow further characterization of torrefied pellets beyond the existing physical-mechanical and chemical parameters already being measured. The method itself is simple enough to be conducted by specialized laboratories and pellet producers alike without great investment cost or training. Even though the impact mill seems to have the most promising features and results, further investigations could be necessary to decide whether the results obtained by this mill type represent the characteristics most relevant to plant operators. To increase the impact of the newly introduced parameter (SGE), it would also be necessary for plant operators to acquire first-hand experience on how the specific grinding energy values translate into adjustment of their mills on-site. Therefore, if the specific grinding energy is to be used as a quality parameter, it might also be useful to complement the single value information with either a particle size distribution curve or the distribution width as part of the characterization.

Supplementary Materials: The following are available online at www.mdpi.com/1996-1073/9/10/794/s1. Figure S1: Picture of the anvil and plunger used during the hardness test, Figure S2: View of the grinding chamber of each of the three mills used, Figure S3: Typical power curve when recoding the power consumption during grinding, Figure S4: Cumulative particle size distribution obtained when using a ball mill on three selected torrefied biomass pellets and four standardized HGI coals, Table S1: Characteristic values of the particle size distributions obtained by the different mills for the different biomass types.

Acknowledgments: The authors would like to thank the managers of the SECTOR-Project for their organizational skills and the European Union for their funding (The SECTOR-Project has received funding from the European Union's Seventh Programme for research, technological development and demonstration under grant agreement n° 282826GA). Furthermore, the authors would like to thank the laboratory staff at the Deutsches Biomasseforschungszentrum who did a great load of the practical work for this study.

Author Contributions: Diana Leistner (D.L.), Nadja Weller (N.W.) and Leilani Darvell (L.I.D.) conceived and designed the experiments while Jan Hari Arti Khalsa (J.H.A.K.) and D.L. did most of the data analyses. J.H.A.K. did most of the writing with input from D.L., N.W. and L.I.D. Ben Dooley (B.D.) and L.I.D. were responsible for the HGI measurements while all other parameters were measured under supervision of D.L. The torrefied biomass pellets were a courtesy of the SECTOR-Project with numerous researchers being involved in the production. These researchers are not included in the list of authors yet their contribution is greatly appreciated.

Conflicts of Interest: The authors declare no conflict of interest.

Abbreviations

BM	ball mill
CM	cutting mill
DU	durability
GM	grinding method
HGI	hardgrove index
HHV	higher heating value
HM	hammer mill
IM	impact mill
PSD	particle size distribution
SGE	specific grinding energy
TF_{temp}	temperature during torrefaction
TF_{time}	duration of torrefaction
TF_{dgr}	degree of torrefaction
TS	technical specification (ISO terminology)

References

1. Directive 2009/28/EC of the European Parliament and of the Council on the Promotion of the Use of Energy from Renewable Sources and Amending and Subsequently Repealing Directives 2001/77/EC and 2003/30/EC. Available online: http://www.ecolex.org/details/legislation/directive-200928ec-of-the-european-parliament-and-of-the-council-on-the-promotion-of-the-use-of-energy-from-renewable-sources-and-amending-and-subsequently-repealing-directives-200177ec-and-200330ec-lex-faoc088009/ (accessed on 17 August 2016).
2. Communication from the Commission to the European Parliament, the Council, the European Economic and Social Committee and the Committee of the Regions Energy Roadmap 2050 /* Com/2011/0885 Final *. Available online: http://eur-lex.europa.eu/legal-content/EN/ALL/?uri=celex%3A52011DC0885 (accessed on 17 August 2016).
3. Mansour, F.A.; Zuwala, J. An evaluation of biomass co-firing in Europe. *Biomass Bioenergy* **2010**, *34*, 620–629. [CrossRef]
4. Sami, M.; Annamalai, K.; Wooldridge, M. Co-firing of coal and biomass fuel blends. *Prog. Energy Combust. Sci.* **2001**, *27*, 171–214. [CrossRef]
5. Loo, S.V.; Koppejan, J. *The Handbook of Biomass Combustion and Co-Firing*; Earthscan: London, UK, 2008.
6. Savolainen, K. Co-firing of biomass in coal-fired utility boilers. *Appl. Energy* **2003**, *74*, 369–381. [CrossRef]
7. Tumuluru, J.S.; Sokhansanj, S.; Hess, J.R.; Wright, C.T.; Boardman, R.D. REVIEW: A review on biomass torrefaction process and product properties for energy applications. *Ind. Biotechnol.* **2011**, *7*, 384–401. [CrossRef]
8. Chew, J.J.; Doshi, V. Recent advances in biomass pretreatment—Torrefaction fundamentals and technology. *Renew. Sustain. Energy Rev.* **2011**, *15*, 4212–4222. [CrossRef]
9. Thrän, D.; Witt, J.; Schaubach, K.; Kiel, J.; Carbo, M.; Maier, J.; Ndibe, C.; Koppejan, J.; Alakangas, E.; Majer, S.; et al. Moving torrefaction towards market introduction—Technical improvements and economic-environmental assessment along the overall torrefaction supply chain through the SECTOR project. *Biomass Bioenergy* **2016**, *89*, 184–200.
10. Koppejan, J.; Sokhansanj, S.; Melin, S.; Madrali, S. *IEA Bioenergy Task 32 Report-Status Overview of Torrefaction Technologies*; Procede Biomass: Enschede, The Netherlands, 2012.
11. Svanberg, M.; Olofsson, I.; Flodén, J.; Nordin, A. Analysing biomass torrefaction supply chain costs. *Bioresour. Technol.* **2013**, *142*, 287–296. [CrossRef] [PubMed]
12. Pirraglia, A.; Gonzalez, R.; Saloni, D.; Denig, J. Technical and economic assessment for the production of torrefied ligno-cellulosic biomass pellets in the US. *Energy Convers. Manag.* **2013**, *66*, 153–164. [CrossRef]
13. Kleinschmidt, C.P. Overview of International Developments in Torrefaction. Available online: http://w.bioenergytrade.org/downloads/grazkleinschmidtpaper2011.pdf (accessed on 17 August 2016).
14. Li, J.; Brzdekiewicz, A.; Yang, W.; Blasiak, W. Co-firing based on biomass torrefaction in a pulverized coal boiler with aim of 100% fuel switching. *Appl. Energy* **2012**, *99*, 344–354. [CrossRef]
15. Biedermann, F.; Brunner, T.; Mandl, C.; Obernberger, I.; Kanzian, W.; Feldmeier, S.; Schwabl, M.; Hartmann, H.; Turowski, P.; Rist, E.; et al. Production of Solid Sustainable Energy Carriers from Biomass by Means of Torrefaction. Available online: https://sector-project.eu/fileadmin/downloads/deliverables/D7.3_7.4-combustion_behaviour__combustion_screening_and_fuel_assessment_Bios__final.pdf (accessed on 1 August 2016).
16. Bridgeman, T.G.; Jones, J.M.; Williams, A.; Waldron, D.J. An investigation of the grindability of two torrefied energy crops. *Fuel* **2010**, *89*, 3911–3918. [CrossRef]
17. Sligar, J. *ACARP the Hardgrove Grindability Index*; ACARP Report I. No. 5; ACARP: Brisbane, QLD, Australia, 1998.
18. *ISO 5074. Hard coal-Determination of Hardgrove Grindability Index*; International Organization for Standardization: Geneva, Switzerland, 1994.
19. Repellin, V.; Govin, A.; Rolland, M.; Guyonnet, R. Energy requirement for fine grinding of torrefied wood. *Biomass Bioenergy* **2010**, *34*, 923–930. [CrossRef]
20. Essendelft, D.T.; Zhou, X.; Kang, B.S.J. Grindability determination of torrefied biomass materials using the Hybrid Work Index. *Fuel* **2013**, *105*, 103–111. [CrossRef]
21. *Impact Grindability Index PMI*; VGB PowerTech: Essen, Germany, 2009.

22. *ISO/PRF TS 17225-8. Solid Biofuels-Fuel Specifications and Classes-Part 8: Graded Thermally Treated and Densified Biomass Fuels*; International Organization for Standardization: Geneva, Switzerland, 2016.

23. Kaliyan, N.; Morey, R. Factors affecting strength and durability of densified biomass products. *Biomass Bioenergy* **2009**, *33*, 337–359. [CrossRef]

24. Joshi, N.R. Relative grindability of bituminous coals on volume basis. *Fuel* **1979**, *58*, 477–478. [CrossRef]

25. Agus, F.; Waters, P.L. Determination of the grindability of coals, shales and other minerals by a modified Hardgrove-machine method. *Fuel* **1971**, *50*, 405–431. [CrossRef]

26. Gucho, E.M.; Shahzad, K.; Bramer, E.A.; Akhtar, N.A.; Brem, G. Experimental study on dry torrefaction of beech wood and miscanthus. *Energies* **2015**, *8*, 3903–3923. [CrossRef]

27. Phanphanich, M.; Mani, S. Impact of torrefaction on the grindability and fuel characteristics of forest biomass. *Bioresour. Technol.* **2011**, *102*, 1246–1253. [CrossRef] [PubMed]

28. Wang, L.; Skreiberg, Ø.; Gronli, M.; Specht, G.P.; Antal, M.J. Is elevated pressure required to achieve a high fixed-carbon yield of charcoal from biomass? Part 2: The importance of particle size. *Energy Fuels* **2013**, *27*, 2146–2156. [CrossRef]

29. Larsson, S.H.; Rudolfsson, M.; Nordwaeger, M.; Olofsson, I.; Samuelsson, R. Effects of moisture content, torrefaction temperature, and die temperature in pilot scale pelletizing of torrefied Norway spruce. *Appl. Energy* **2013**, *102*, 827–832. [CrossRef]

30. Järvinen, T.; Agar, D. Experimentally determined storage and handling properties of fuel pellets made from torrefied whole-tree pine chips, logging residues and beech stem wood. *Fuel* **2014**, *129*, 330–339. [CrossRef]

31. *ISO 17225-2. Solid Biofuels-Fuel Specifications and Classes-Part 2: Graded Wood Pellets*; International Organization for Standardization: Geneva, Switzerland, 2014.

32. *ISO 17225-6. Solid Biofuels-Fuel Specifications and Classes-Part 6: Graded Non-Woody Pellets*; International Organization for Standardization: Geneva, Switzerland, 2014.

33. Li, H.; Liu, X.; Legros, R.; Bi, X.T.; Lim, C.J.; Sokhansanj, S. Pelletization of torrefied sawdust and properties of torrefied pellets. *Appl. Energy* **2012**, *93*, 680–685. [CrossRef]

34. Shang, L.; Ahrenfeldt, J.; Holm, J.K.; Sanadi, A.R.; Barsberg, S.; Thomsen, T.; Stelte, W.; Henriksen, U.B. Changes of chemical and mechanical behavior of torrefied wheat straw. *Biomass Bioenergy* **2012**, *40*, 63–70. [CrossRef]

35. Shang, L.; Nielsen, N.K.; Dahl, J.; Stelte, W.; Ahrenfeldt, J.; Holm, J.K.; Thomsen, T.; Henriksen, U.B. Quality effects caused by torrefaction of pellets made from Scots pine. *Fuel Process. Technol.* **2012**, *101*, 23–28. [CrossRef]

Article

Effects of Syngas Cooling and Biomass Filter Medium on Tar Removal

Sunil Thapa [1,2], Prakashbhai R. Bhoi [2,3], Ajay Kumar [2,3,*] and Raymond L. Huhnke [2,3]

[1]   Environmental Science Graduate Program, Oklahoma State University, Stillwater, OK 74078, USA;
      sunil.thapa@okstate.edu
[2]   Biobased Products and Energy Center, Oklahoma State University, Stillwater, OK 74078, USA;
      prakash.bhoi@okstate.edu (P.R.B.); raymond.huhnke@okstate.edu (R.L.H.)
[3]   Biosystems and Agricultural Engineering, Oklahoma State University, Stillwater, OK 74078, USA
*   Correspondence: ajay.kumar@okstate.edu; Tel.: +1-405-744-8396; Fax: +1-405-744-6059

Academic Editor: Jaya Shankar Tumuluru
Received: 19 December 2016; Accepted: 6 March 2017; Published: 11 March 2017

Abstract: Biomass gasification is a proven technology; however, one of the major obstacles in using product syngas for electric power generation and biofuels is the removal of tar. The purpose of this research was to develop and evaluate effectiveness of tar removal methods by cooling the syngas and using wood shavings as filtering media. The performance of the wood shavings filter equipped with an oil bubbler and heat exchanger as cooling systems was tested using tar-laden syngas generated from a 20-kW downdraft gasifier. The tar reduction efficiencies of wood shavings filter, wood shavings filter with heat exchanger, and wood shavings filter with oil bubbler were 10%, 61%, and 97%, respectively.

Keywords: biomass; gasification; tar; syngas cleaning; dry filter

---

1. Introduction

Conventional fuels such as coal, oil, and natural gas have significant negative environmental impacts and cannot be used on sustainable basis. These concerns have helped shift energy consumption towards renewable and environment friendly sources like biomass-derived energy [1]. Energy from biomass materials are obtained either by direct burning or conversion into liquid or gaseous fuels for practical applications. Technologies such as combustion, gasification, pyrolysis, hydrothermal liquefaction, and fermentation are successfully developed, while few are either commercialized or are under development that convert biomass into useful forms of energy. Among these technologies, gasification offers flexibility in feedstock selection (e.g., agricultural and forest residues, byproducts of food industry and bio-refineries, organic municipal waste), and the product gases (syngas) can be used as fuel to produce heat and electricity. Hence, the gasification is considered one of the most promising technologies [2,3] that utilizes biomass for energy production [4]. Compared to direct combustion, gasification has distinct advantages, such as increased power generation efficiency (up to a 60% increase) and the ability to use the syngas for products other than electricity [5,6].

Biomass gasification is a proven technology and converts carbonaceous materials into syngas, which consists of hydrogen (H_2), carbon monoxide (CO), and carbon dioxide (CO_2) with methane (CH_4), water (H_2O), and higher hydrocarbons (HC) in minor quantities [4,7]. The application of syngas includes electricity generation using internal combustion (IC) engines and gas turbines and production of chemicals and liquid fuels [8]. Regardless of many advantages of biomass gasification and wide application of syngas, commercial acceptability of the technology still faces challenges due to difficulty in cleaning the produced syngas [8]. Impurities—such as tars, particulate matters, sulfur, and ammonia—are produced during gasification and entrained in the syngas. In most applications, these impurities must be removed before using syngas. Among these contaminants, tar is a big challenge.

Tar forms during condensation of syngas, and is classified as primary, secondary, and tertiary. Formation of primary tars, in the pyrolysis step, occurs at low temperatures (below 500 °C). In the oxidation step, with increase in temperature (above 500 °C), the primary tars transform and begin to rearrange as secondary tars. Further increase in temperature (above 800 °C) leads to formation of tertiary tars [4]. Tar consists of a wide range of hydrocarbons that can cause blockage in the downstream equipment, engine, and compressors [9–11]. The composition and quantity of tar in syngas depends upon gasification conditions and feedstock properties [2]. The acceptable limit varies depending on the application as shown in Table 1 [9–12].

Table 1. Tar acceptance limit for different devices. IC: internal combustion.

End Use Device	Tar Acceptance Limit (g/Nm^3)
Industrial gas turbine	<0.005
Fuel cells	<0.001
IC and diesel engines	0.01–0.1
Compressors	0.05–0.5

Removal of tar from the syngas is done either by chemical methods, such as catalytic and thermal cracking [5,13,14], which are expensive to operate [5,15] or by physical methods, such as wet scrubbers and water spray, which have issues with disposing tar mixed solvents [15–17]. Besides disposal of tar-mixed solvent, major disadvantages of using a wet cleaning system include decrease in heating value of the producer gas and the net energy efficiency. Therefore, development of tar removal technology with low cost, minimal disposal issues, and high efficiency are instrumental for future of gasification-based technologies [5,16]. Biomass-based dry filters could be an economical and environmental friendly option. Biomass-based filters using corn cobs, wood shavings, and dry coconut coir, charcoal and saw dust have been explored [17–19]. Corn cobs and woodchips reduced the tar content to 2 g/Nm^3 [18]. To further improve tar removal, other studies have used forms of direct cooling system, including wet scrubbers and water spray towers, before a dry biomass filter [15–17]. However, wet cleaning systems require expensive waste water treatment, so safe discharge of the tar mixed solution is a challenge [15–17]. A series of heat exchangers [20–22] and oil scrubber [23–27] have shown to be effective in reducing tar to as low as 10 mg/Nm^3. The tar collected from the heat exchanger and used oil from the oil scrubber can be recycled back to the gasifier [28,29], which eliminates waste effluent treatment. A cleaning system equipped with an indirect cooling system that eliminates waste effluent treatment process and filter mediums which can be reused as gasifier feedstock is a promising alternative to conventional cleaning technologies. However, dry biomass-based filter systems are not tested in combination with indirect cooling system (heat exchanger) or oil scrubber.

The goal of this study was to improve tar removal efficiencies of a dry biomass-based filter in combination with a heat exchanger and oil bubbler. The tar removal efficiencies were evaluated with three filtering methods: (i) wood shavings filter; (ii) wood shavings filter after cooling with a heat exchanger; and (iii) wood shavings filter after cooling with an oil bubbler.

2. Materials and Methods

2.1. Experimental Set Up

Schematic diagrams of the experimental set up are shown in Figures 1 and 2. Oklahoma State University patented downdraft gasifier (20 kW) [30] with an average feedstock consumption rate of 15 kg/h was operated at a low equivalence ratio (0.20) to generate syngas with high tar content. Operational conditions of gasifiers for each treatment are presented in Table 2. Switchgrass used as gasification feedstock was grown at the Agronomy Research Station of Oklahoma State University

(Stillwater, OK, USA). A 10-layered cylindrical biomass filter system with total surface area of 1.26 m^2 (bed height of 0.96 m and diameter of 0.17 m) was designed. Syngas entered into the filter at the bottom at a flow rate of 35 Nm^3/h. Syngas exited through the filter top (Figure 3). Micro data acquisition (DAQ) K-type thermocouples (Omega Engineering, Norwalk, CT, US.) were installed to measure the gas temperature at the inlet and outlet of the filter system. The pressure drop across the filter system was measured using U-tube manometers (Dwywe Inc, Wilmington. NC, US).

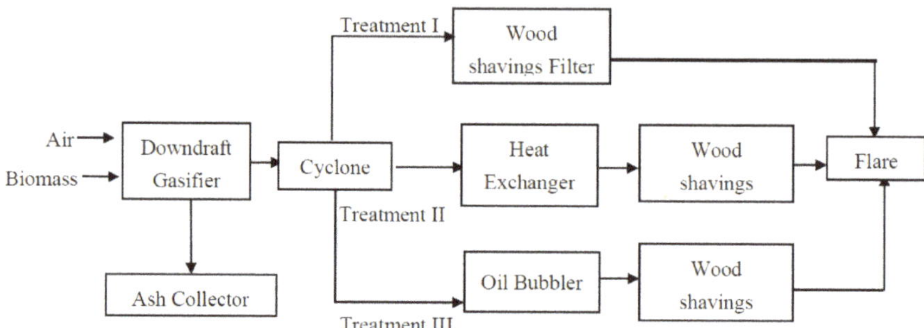

Figure 1. Schematic diagram of the three gas cleaning treatments: (**I**) wood shavings filter; (**II**) heat exchanger + wood shavings filter; and (**III**) oil bubbler + wood shavings filter.

Figure 2. Wood shavings filter column installed with a gasifier system at Oklahoma State University.

Table 2. Gasifier operating conditions.

Parameters	Treatment I	Treatment II	Treatment III
Fuel feed rate, kg/h	16.5	16	16.5
Combustion zone temperature, °C	719	700	703
Equivalence ratio	0.20	0.20	0.20

Figure 3. 10-layered biomass-based filter system.

The wood shavings, screened through 2 mm wire mesh, with a particle size between 2 mm and 1.5 cm were used as filter medium (Figure 4). Wood shavings were purchased from a local sawmill in Stillwater. A total of 4.0 kg of wood shavings was used for each run to maintain the uniform packed bed density of the filter. Syngas was passed through the filter for 1.0 h after the gasifier was at equilibrium condition. Chilled water (5 °C, flow rate of 15 L/min) for single tube heat exchanger (1.19 m height and 0.02 m diameter; Figure 5) was supplied by a water chiller (Model 30070, Schreiber Engineering Corporation, Cerritos, CA, USA). The oil bubbler (0.48 m height, 0.38 m diameter and 10 mm bubble size with 1-inch oil level; Figure 6) used canola oil that was supplied by Jedwards International, Inc. (Quincy, MA, USA). The oil bubbler was also cooled with chilled water (5 °C, flow rate of 15 L/min). The inlet and outlet syngas tar contents, pressure drop across the filter, temperatures at inlet and outlet of the filter, heating values of syngas and gas temperatures at inlet and outlet of the heat exchanger were recorded.

Figure 4. Wood shavings used as the filter medium for all treatments.

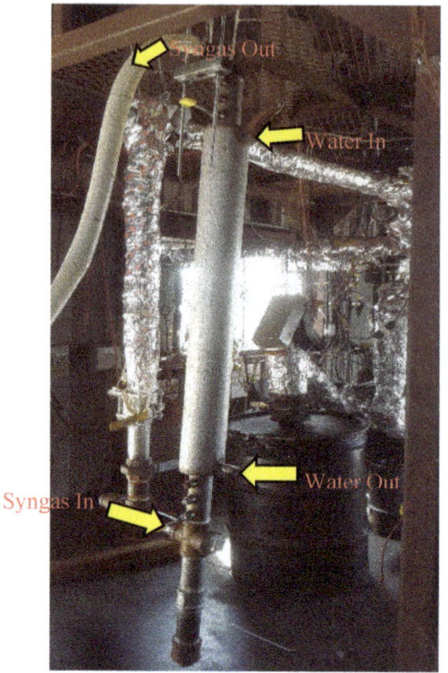

Figure 5. A single tube heat exchanger used for cooling syngas for treatment II.

Figure 6. Oil bubbler unit for treatment III.

2.2. Tar and Syngas Sampling and Analysis

The tar removal efficiency of the filter was determined by comparing tar content at inlet and outlet of the filter. Tar sampling protocol by Good et al. [31] was used. The method consists of a series of six impingement bottles in which the syngas was passed. The first one served as a moisture collector. The gas then flowed through a series of four impinger bottles filled with a solvent, i.e., acetone to dissolve the tars. The last bottle was kept empty to ensure the collection of final condensates. For tar analysis, syngas was sampled at the inlet of cleaning system approximately 30 min after the gasifier reached equilibrium condition. The gas at the filter outlet was sampled 30 min after start of filter operation. During each test, syngas was separately sampled every 20 min at the filter outlet for analysis of gas composition using a gas chromatographer (Model CP-3800, Varian, Inc., Palo Alto, CA, USA). The tar content in syngas was calculated as the ratio of the weight of tar in sampled gas (g) and volume of sampled gas (m^3).

3. Results and Discussion

Treatments were designed to investigate tar reduction (removal) efficiencies of wood shavings filter (dry filter) with and without syngas cooling and bubbling systems. Accordingly, the tar contents were measured at the inlet and outlet of the cleaning systems for the three treatments: (i) wood shavings filter; (ii) wood shavings filter after a heat exchanger; and (iii) wood shavings filter after an oil bubbler. Pressure drop, an important indicator in determining the tar absorbed by filter medium, was measured across the wood shavings filter over time. Similarly, because the temperature of syngas plays a vital role in condensation and deposition of tar, temperatures were recorded at the inlet and outlet of the cleaning systems. The gasifier was purposefully operated at low temperatures, which resulted in high syngas tar content to test the capabilities of the cleaning systems. The removal of tar can involve condensation or absorption process. In this experiment, the removal of tar by wood shavings involves condensation. Whereas, the removal of tar by oil is an absorption-based process because the solvent (oil) dissolves the tar products.

3.1. Tar Reduction Efficiencies

3.1.1. Tar Reduction Efficiency of Wood Shavings Filter (Treatment I)

Tar content at the inlet and outlet of wood shavings filter were 78 g/Nm^3 and 70 g/Nm^3, respectively (Table 3). A study [18] using corn cobs as the filter medium reported outlet tar content of 2 g/Nm^3. The lower outlet tar content using corn cobs may be attributed to the two layers of sponges used in combination with the corn cobs. The tar removal efficiency of the wood shavings filter obtained in this study (10%) was much lower compared to 97% reported using a water scrubber and coconut coir filter [17] and 94% using spray towers and wood shavings filter [19]. Observation of biomass filter layers (Treatment I in Figure 7) revealed that more tar was adsorbed at the upper section and around the periphery of the filter. This was attributed to the lower temperatures near the periphery of the filter.

Table 3. Tar contents of raw and cleaned syngas (dry basis) with the three tar cleaning treatments.

Treatment	Raw Gas Tar Concentration (g/Nm^3)	Cleaned Gas Tar Concentration (g/Nm^3)
Wood shavings filter (I)	78	70
Heat exchanger + wood shavings filter (II)	70	27
Oil bubbler + wood shavings filter (III)	70	1.9

Figure 7. *Cont.*

Bottom	Bottom	Bottom
Treatment I	Treatment II	Treatment III

Figure 7. Top and bottom sections of the filter for three treatments: (**I**) wood shavings filter; (**II**) heat exchanger + wood shavings filter; and (**III**) oil bubbler + wood shavings filter.

3.1.2. Tar Reduction Efficiency of Wood Shavings Filter in Combination with Heat Exchanger (Treatment II)

The inlet and outlet syngas tar contents of the filter system equipped with a wood shavings filter and heat exchanger, were 70 g/Nm^3 and 27 g/Nm^3, respectively, resulting in tar removal efficiency of 61% (Table 3). Unlike previous studies [15,17,19] that used direct (contact) cooling systems, such as spray towers and water scrubbers; the heat exchanger used in this study was an indirect (no contact) cooling system. Compared to Treatment I (only wood shavings filter), indirect cooling of syngas using a heat exchanger (Treatment II), reduced the dew point of tar and significantly improved tar removal efficiency. Figure 7 presents the pictorial comparison of tar adsorbed at top and bottom sections of the wood shavings filter. Similar to the observation made in Treatment I, more tar deposited at the top section and around periphery of the filter in Treatment II. Figure 7 also shows that the addition of the heat exchanger clearly aided in deposition of more tar on the wood shavings filter. However, the final tar content (27 g/Nm^3) obtained in this treatment was higher than those obtained with dry filter used with direct (touch) syngas cooling and cleaning systems, such as sand bed filter with water spray towers (1.5 g/Nm^3 final tar content) [15] and coconut coir filter with water scrubber (1.4 g/Nm^3 final tar content) [17]. Similarly, the final tar content of this treatment is high than those of cleaning systems with a heat exchanger with bag house filter (35 mg/Nm^3) [21]; and heat exchanger with venture scrubber (10 mg/Nm^3) [22]. High outlet tar content in this study can be attributed to two primary reasons: (i) indirect cooling, used in this study, has low tar removal effectiveness; and (ii) inlet tar content was high (70 g/Nm^3 vs. <10 g/Nm^3) [15,19]. The result from Treatment II is comparable to the tar removal efficiency (61%) of the system with a cyclone, heat exchanger and oil bath filter [32]. These result, shows that indirect cooling of syngas with the heat exchanger increased the tar removal efficiency of the dry filter but further cleaning is still required if the syngas is used in an Internal Combustion (IC) engine. However, since water and some tar are condensed in the heat exchanger, the design of the heat exchanger must allow water and tars to flow out of the heat exchanger pipes. Use of a single vertical tube allowed us to collect water and tar condensed during the test. Tar deposition along the inner surface of the heat exchanger was minimal but after several runs, the inner surface required cleaning.

3.1.3. Tar Reduction Efficiency of Wood Shavings Filter with Vegetable Oil Bubbler (Treatment III)

To investigate the tar removal efficiency of wood shavings filter equipped with vegetable oil bubbler as cooling unit, the cleaning system was installed as specified in Figure 1 (Treatment III). Tar content at the inlet and outlet of the cleaning system were observed as 70 g/Nm^3 and 1.9 g/Nm^3, respectively (Table 3). The tar content at inlet of wood shavings filter (after oil bubbler) was 3.8 g/Nm^3, which suggests that a large portion of the tar was condensed in oil bubbler (as shown in Figure 8), and is also evident from Figure 4, which shows that very little tar was condensed on the wood shavings filter.

Figure 8. Canola oil (used in oil bubbler for treatment III) before and after the test.

High tar removal efficiency of the cleaning system equipped with oil bubbler (97%) indicates that, unlike in previous studies [24,25,27] that have used oil scrubber, an oil bubbler is also effective for removal of syngas tar. High tar removal efficiency of 98% with sunflower oil [24] and final tar content of 0.022 g/Nm^3 with waste palm cooking oil [27] have been reported. The high tar removal efficiency of the oil-based cleaning system is attributed to oil's lipophilicity characteristics, the ability of the oil to dissolve non-polar hydrocarbons [27]. Tar compounds are lipophilic in nature and can mix well with vegetable oils as these oils have saturated and unsaturated fatty acids.

Feasibility of removing syngas tar with biomass and oil is promising because the tar mixed oil and wood shavings can be reused in gasifier reactors as feedstock, eliminating the need to treat waste effluent. For example, the scrubbing oil was reused and tars were recycled into the gasifier by the Energy Research Center of the Netherlands [28]. Oil used in removing tar was put in the regeneration process using filtration and centrifugal sedimentation techniques and reused in the scrubber [29]. Wood shavings also removed syngas moisture depicted by the high moisture content of the filter after the test. However, the tar content at the outlet of oil bubbler was still not low enough for engine application. In conclusion, oil may have been effective in reduction of heavy tar [24], but additional cleaning is necessary for the removal of light tars.

3.2. Variation of Pressure Drop across Wood Shavings Filter for the Three Treatments

Pressure drop across the wood shavings filter depends on amount of tar accumulated in the filter medium. As shown in Figure 9, pressure drop across the filter increased with time due to continued accumulation of tar on the filter medium for all three treatments. However, throughout the test duration, pressure drop was the highest for Treatment II (when heat exchanger was used before the filter) and the lowest for Treatment III (when oil bubbler was used before the filter). The trend of pressure drop indicated that tar deposition on the filter was the highest for Treatment II and the lowest for Treatment III. However, overall tar removal efficiency was the highest for Treatment III (97%) and the lowest for Treatment I (10%). These observations indicated that the highest tar removal for Treatment III is due to the oil bubbler, which removed most of the tars leaving only a small quantity of tar deposited on the filter hence the pressure drop across the filter was minimal. Syngas tar content measured at the outlet of the oil bubbler (3.8 g/Nm^3) confirmed the finding that only a small quantity of tar (3.8 g/Nm^3) was removed by the filter medium and most of the tar was removed by the oil bubbler (66.2 g/Nm^3). For Treatment I, as the filter medium was not augmented with any other cleaning method, the pressure drop increased with increasing tar accumulation on the filter over time. In Treatment II, use of a heat exchanger before the filter medium reduced the temperature of syngas entering into the filter medium from 135 °C to 71 °C, which led to increased condensation of the tar on the filter medium. Tar deposition observed on the heat exchanger was minimal, indicating that most of the tar was removed by the filter. Higher pressure drop across the filter for Treatment II as compared to Treatment I also indicated that tar deposition on the filter for Treatment II (with heat exchanger) was higher than that for Treatment I (only filter). Hence, the use of heat exchanger reduced the syngas temperature and increased tar deposition on the filter medium. The pressure drops across the filter for all three treatments (0.2–0.5 in of H_2O) were lower compared to those reported by others (0.5–2 in of H_2O) [15,17] due to low condensation of tar by wood shavings. This low pressure drop is beneficial for power generation application because pressure available at the engine manifold is high and prevents high back pressure in the gasifier.

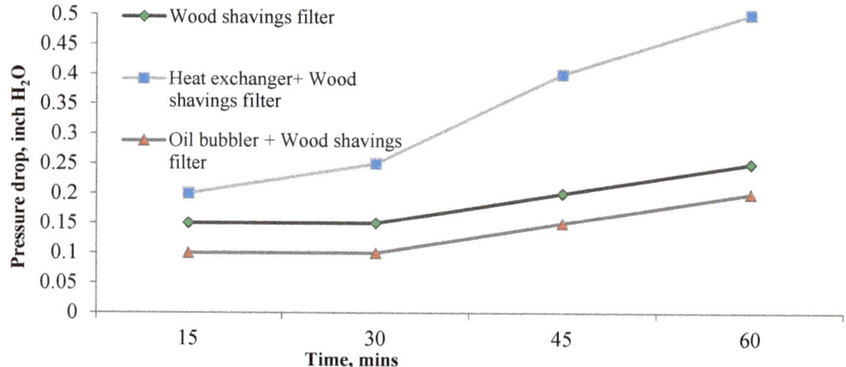

Figure 9. Variation of pressure drop of wood shavings filters over time.

3.3. Variation of Gas Temperature at Inlet and Outlet of Wood Shavings Filter for Three Treatments

Temperature of syngas is a key parameter affecting condensation of tar. Figure 10 shows the 1 h average syngas temperature at the inlet and outlet of the wood shavings filter for the three treatments. The filter inlet temperature for Treatment I was the highest, followed by Treatment II and Treatment III. The trend shows that the cooling by the heat exchanger and the oil bubbler were effective. The syngas temperature entering into the cleaning system was the same (about 135 °C) for all three treatments; however, because of the use of heat exchanger and oil bubbler before the filter, the syngas temperature at the filter inlet was different. The average syngas temperature entering into the cleaning system was low (about 135 °C), due to the use of long piping between the cyclone and cleaning system. As a result, tar condensation could have happened along the piping. The difference between inlet and outlet syngas temperatures of the filter (Figure 10) was the highest for Treatment I, followed by Treatments II and III. This trend can be attributed to the effectiveness of heat exchanger and oil bubbler in reducing gas temperature at the filter inlet.

As expected, low syngas temperatures at the filter led to high condensation on the wood shavings medium. In addition, low temperature of 30 °C is desired for feeding into IC engine, [10]. The oil bubbler (Treatment III) was effective in reducing syngas temperature at the filter outlet to 27 °C. The heat exchanger-based cooling system also reduced the syngas temperature; however, the outlet syngas temperature (58 °C) was still higher than desired for most engine applications. Using a heat exchanger with multiple tubes may further reduce the temperature sufficient for engine applications.

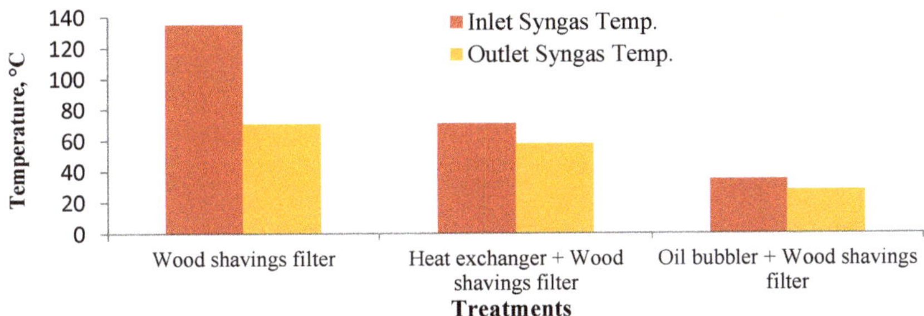

Figure 10. Syngas temperature at the inlet and outlet of the wood shavings filter for the three treatments.

3.4. Heating Value of Syngas

Heating value of syngas affects performance of the downstream power generation unit [21]. The lower heating values (LHVs) of syngas sampled at the outlet of the wood shavings filter for the three treatments were in the range of 5–6 MJ·Nm^{-3}. These results are comparable to 5.3 MJ·Nm^{-3} produced from a downdraft gasifier with olive kernel as feedstock and wet scrubber and heat exchanger as a cleaning unit [22] and 5.79 MJ·Nm^{-3} produced from an 18 kW gasifier using hardwood chips as feedstock [33]. The average gas composition for each of the three treatments are presented in Table 4. The variation in heating values of product gas can be attributed to the difference in composition of combustible gases, such as H_2, CO and CH_4.

Table 4. Syngas composition and heating value for the three cleaning treatments.

Gas components	Treatment I	Treatment II	Treatment III
H_2 (% v/v)	10.2	12.7	11.5
N_2 (% v/v)	52.7	50.3	50.0
O_2 (% v/v)	2.6	2.0	1.5
CO (% v/v)	12.8	14.0	14.6
CH_4 (% v/v)	2.9	2.9	4.8
CO_2 (% v/v)	15.6	14.2	19.1
LHV (MJ·Nm^{-3})	5.432	5.766	5.680

4. Conclusions

The performance of three types of syngas cleaning systems using wood shavings as the filter medium was evaluated. Tar removal efficiencies of the three treatments were: wood shavings filter (10.28%) < wood shavings filter with heat exchanger (60.30%) < wood shavings filter with oil bubbler (97%). Even though the heat exchanger reduced the syngas temperature and led to increased condensation of tars, the vegetable oil bubbler was more effective in the removal of tars because of the oil's ability to absorb tar. Tar deposited at the top and around the periphery of the wood shavings filter was the highest when the heat exchanger was used.

Acknowledgments: The authors would like to acknowledge Oklahoma State University Research Foundation and Oklahoma Agricultural Experiment Station for their support.

Author Contributions: All authors significantly contributed to the scientific study and writing. Thapa performed tests and collected data, whereas, Bhoi, Kumar and Huhnke supervised the project, and all authors contributed to writing the manuscript.

Conflicts of Interest: The authors declare no conflict of interest.

References

1. Srirangan, K.; Akawi, L.; Moo-Young, M.; Chou, C.P. Towards sustainable production of clean energy carriers from biomass resources. *Appl. Energy* **2012**, *100*, 172–186. [CrossRef]
2. Devi, L.; Ptasinski, K.J.; Janssen, F.J. A review of the primary measures for tar elimination in biomass gasification processes. *Biomass Bioenergy* **2003**, *24*, 125–140. [CrossRef]
3. Heidenreich, S.; Foscolo, P.U. New concepts in biomass gasification. *Prog. Energy Combust. Sci.* **2015**, *46*, 72–95. [CrossRef]
4. Molino, A.; Chainese, S.; Musmarra, D. Biomass gasification technology: The state of the art overview. *J. Energy Chem.* **2016**, *25*, 10–25. [CrossRef]
5. Anis, S.; Zainal, Z.A. Tar reduction in biomass producer gas via mechanical, catalytic and thermal methods: A review. *Renew. Sustain. Energy Rev.* **2011**, *15*, 2355–2377. [CrossRef]
6. Advantages and Efficiency of Gasification. Available online: https://www.netl.doe.gov/research/coal/energy-systems/gasification/gasifipedia/clean-power (accessed on 31 January 2017).
7. Ahmad, A.A.; Zawawi, N.A.; Kasim, F.H.; Inayat, A.; Khasri, A. Assessing the gasification performance of biomass: A review on biomass gasification process conditions, optimization and economic evaluation. *Renew. Sustain. Energy Rev.* **2016**, *53*, 1333–1347. [CrossRef]
8. Huang, J.; Schmidt, K.G.; Bian, Z. Removal and conversion of tar in syngas from woody biomas gasification for power utilization using catalytic hydrocracking. *Energies* **2011**, *4*, 1163–1177. [CrossRef]
9. Asadullah, M. Biomass gasification gas cleaning for downstream applications: A comparative critical review. *Renew. Sustain. Energy Rev.* **2014**, *40*, 118–132. [CrossRef]
10. Baratieri, M.; Baggio, P.; Bosio, B.; Grigiante, M.; Longo, G.A. The use of biomass syngas in IC engines and CCGT plants: A comparative analysis. *Appl. Therm. Eng.* **2009**, *29*, 3309–3318. [CrossRef]
11. Singh, R.N.; Mandovra, S.; Balwanshi, J. Performance of evaluation of "jacketed cyclone" for reduction of tar from producer gas. *Int. Agric. Eng. J.* **2013**, *22*, 1–5.
12. Hasler, P.; Nussbaumer, T. Gas cleaning for IC engine applications from fixed bed biomass gasification. *Biomass Bioenergy* **1999**, *16*, 385–395. [CrossRef]
13. Chainese, S.; Loipersbock, J.; Malits, M.; Rauch, R.; Hofbauer, H.; Molino, A.; Musmarra, D. Hydrogen from the high temperature water gas shift reaction with an industrial Fe/Cr catalyst using biomass gasification tar rich synthesis gas. *Fuel Process. Technol.* **2015**, *132*, 39–48. [CrossRef]
14. Chainese, S.; Fail, S.; Binder, M.; Rauch, R.; Hofbauer, H.; Molino, A.; Blasi, A.; Musmarra, D. Experimental investigations of hydrogen production from CO catalytic conversion of tar rich syngas by biomass gasification. *Catal. Today* **2016**, *277*, 182–191. [CrossRef]
15. Pathak, B.S.; Kapatel, D.; Bhoi, P.R.; Sharma, A.M.; Vyas, D.K. Design and development of sand bed filter for upgrading producer gas to IC engine quality fuel. *Int. Energy J.* **2007**, *8*, 15–20.
16. Nakamura, S.; Unyaphan, S.; Yoshikawa, K.; Kitano, S.; Kimura, S.; Shimizu, H.; Tiara, K. Tar removal performance of bio-oil scrubber for biomass gasification. *Biofuels* **2014**, *5*, 597–606. [CrossRef]
17. Rameshkumar, R.; Mayilsamy, K. A novel compact bio-filter system for a downdraft gasifier: An experimental study. *AASRI Procedia* **2012**, *3*, 700–706. [CrossRef]
18. Allesina, G.; Pedrazzi, S.; Montermini, L.; Giorgini, L.; Bortolani, G.; Tartarin, P. Porous filtering media comparison through wet and dry sampling of fixed bed gasification. *J. Phys. Conf. Series* **2014**, *547*. [CrossRef]
19. Pareek, D.; Joshi, A.; Narnaware, S.; Verma, V.K. Operational experience of agro-residue briquettes based power generation system of 100 kW capacity. *Int. J. Renew. Energy Res.* **2012**, *2*, 477–485.
20. Henriken, U.; Ahrenfeldt, J.; Jensen, T.K.; Gobel, B.; Bentzen, J.D.; Hindsgaul, C.; Sorensen, L.H. The design, construction and operation of a 75 kW two-stage gasifier. *Energy* **2006**, *31*, 1542–1553. [CrossRef]
21. Raman, P.; Ram, N.K.; Gupta, R. A dual fired downdraft gasifier system to produce cleaner gas for power generation: Design, development and performance analysis. *Energy* **2013**, *54*, 302–314. [CrossRef]
22. Margaritis, N.K.; Grammelis, P.; Vera, D.; Jurado, F. Assessment of operational results of a downdraft biomass gasifier coupled with a gas engine. *Proc. Soc. Behav. Sci.* **2012**, *48*, 857–867. [CrossRef]
23. Phuphukrat, T.; Namioka, T.; Yoshikawa, K. Absorptive removal of biomass tar using water and oily materials. *Bioresour. Technol.* **2011**, *102*, 543–549. [CrossRef]

24. Paethanom, A.; Bartocci, P.; Alessandro, B.D.; Amico, M.D.; Testarmata, F.; Moriconi, N.; Slopiecka, K.; Yoshikawa, K.; Fantozzi, F. A low-cost pyrogas cleaning system for power generation: Scaling up from lab to pilot. *Appl. Energy* **2013**, *111*, 1080–1088. [CrossRef]

25. Bhoi, P.R.; Huhnke, R.L.; Kumar, A.; Payton, M.E.; Patil, K.N.; Whiteley, J.R. Vegetable oil as a solvent for removing producer gas tar compounds. *Fuel Proc. Technol.* **2015**, *133*, 97–104. [CrossRef]

26. Nakamura, S.; Siriwat, U.; Yoshikawa, K.; Kitano, S. Development of tar removal technologies for biomass gasification using the byproducts. *Energy Procedia* **2015**, *75*, 208–213. [CrossRef]

27. Ahmad, N.A.; Zainal, Z.A. Performance and chemical composition of waste palm cooking oil as scrubbing medium for tar removal from biomass producer gas. *J. Natural Gas Sci. Eng.* **2016**, *31*, 256–261. [CrossRef]

28. MILENA and OLGA Get together for High Efficiency and Low Tar. Available online: http://www.modernpowersystems.com/features/featuremilena-and-olga-get-together-for-high-efficiency-and-low-tar-4815584/ (accessed on 12 December 2016).

29. Tarnpradab, T.; Unyaphan, S.; Takahasi, F.; Yoshikawa, K. Improvement of the biomass tar removal capacity of scrubbing oil regenerated by mechanical soild-liquid separation. *Energy Fuels* **2017**, *31*, 1564–1573. [CrossRef]

30. Patil, K.N.; Huhnke, R.L.; Bellmer, D. Downdraft Gasifier with Internal Cyclonic Combustion Chamber. U.S. Patent 8,657,892 B, 25 February 2014.

31. Good, J.; Ventress, L.; Knoef, H.; Zielke, U.; Hansen, P.L.; Kamp, W.V.D.; Wild, P.D.; Coda, B.; Paasen, S.V.; Kiel, J. *Sampling and Analysis of Tar and Particles in Biomass Producer Gases*; European Committee for Standardization (CEN): Brussels, Belgium, 2005.

32. Atnaw, S.M.; Kueh, S.C.; Sulaiman, S.A. Study on tar generated from downdraft gasification of oil palm fronds. *Sci. World J.* **2014**. [CrossRef] [PubMed]

33. Shah, A.; Srinivasan, R.; To, S.D.; Columbus, E.P. Performance and emissions of a spark-ignited engine driven generator on biomass based syngas. *Bioresour. Technol.* **2010**, *101*, 4656–4661. [CrossRef] [PubMed]

Brief Report

Pyrolysis Kinetics of the Arid Land Biomass Halophyte Salicornia Bigelovii and Phoenix Dactylifera Using Thermogravimetric Analysis

Prosper Dzidzienyo [1], Juan-Rodrigo Bastidas-Oyanedel [1,2,*] and Jens Ejbye Schmidt [1,2]

[1] Department of Chemical Engineering, Khalifa University of Science and Technology, Masdar City Campus, P.O. Box 54224, Abu Dhabi, United Arab Emirates; dmpeekay@gmail.com (P.D.); jschmidt@masdar.ac.ae (J.E.S.)

[2] Chemistry Department, Khalifa University of Science and Technology, Masdar City Campus, P.O. Box 54224, Abu Dhabi, United Arab Emirates

* Correspondence: yanauta@gmail.com; Tel.: +97-150-502-8842

Received: 6 May 2018; Accepted: 16 July 2018; Published: 30 August 2018

Abstract: Biomass availability in arid regions is challenging due to limited arable land and lack of fresh water. In this study, we focus on pyrolysis of two biomasses that are typically abundant agricultural biomasses in arid regions, focusing on understanding the reaction rates and Arrhenius kinetic parameters that describe the pyrolysis reactions of halophyte *Salicornia bigelovii*, date palm (*Phoenix dactylifera*) and co-pyrolysis biomass using thermo-gravimetric analysis under non-isothermal conditions. The mass loss data obtained from thermogravimetric analysis of *S. bigelovii* and date palm revealed the reaction rate peaked between 592 K and 612 K for *P. dactylifera* leaves and 588 K and 609 K for *S. bigelovii* at heating rates, 5 K/min, 10 K/min and 15 K/min during the active pyrolysis phase. The activation energy for *S. bigelovii* and *P. dactylifera* leaves during this active pyrolysis phase were estimated using the Kissinger method as 147.6 KJ/mol and 164.7 KJ/mol respectively with pre-exponential factors of 3.13×10^9/min and 9.55×10^{10}/min for the respective biomasses. Other isoconversional models such as the Flynn-Wall-Ozawa were used to determine these kinetic parameters during other phases of the pyrolysis reaction and gave similar results.

Keywords: halophytes; *Phoenix dactylifera*; *Salicornia bigelovii*; thermogravimetric analysis

---

1. Introduction

Any arid region biomass-based renewable energy form has to confront a number of challenges, such as the unavailability of enough arable land and lack of fresh water [1]. The two biomasses considered in this study were halophyte *Salicornia bigelovii*, which can be grown in arid lands using saltwater [2] and date palms (*Phoenix dactylifera*), a native arid land biomass and one of the most abundant agricultural residues in arid regions. Pyrolysis of biomass was identified as an effective way of producing pre-cursors for jet-fuel production. Pyrolysis of biomass yields pyro-oils, pyro-char and gases in varying proportions depending on process parameters [3–5]. Pyro oil yields of up to 35% have been obtained by various authors [6] while fast pyrolysis yields have reached up to 75% [5]. Other authors have studied the effects of temperature on pyrolysis oil yields [7,8].

Arid-land lignocellulose biomass pyrolysis literature is scarce [9–11], and co-pyrolysis of these types of biomass have not been reported before. Both *S. bigelovii* and date palm have attracted attention in the biorefinery of arid-land biomass [1,12,13]. From the sustainability perspective, both plants are addapted to the harsh conditions of arid-land. *S. bigelovii* can grow in arid-land coastal areas using seawater, thereby reducing the stress on freshwater demand [12,13]. Date palm leaf residues', including leaflets and rachis, annual global production is estimated to be over 6 million tonnes [1]. Hence, the pyrolysis of this biomass is an opportunity to divert these residues from landfilling.

Understanding the kinetic parameters and effect of heating rates on reaction rate at different stages in the pyrolysis process is important in the design of pyrolysis reactors specifically for the investigated biomasses and in scaling up the process. It is also necessary in predicting the extent of reaction under different experimental conditions at different times. This work studies the pyrolysis reactions using thermogravimetric analysis for *S. bigelovii*, date palm, and co-pyrolysis of both at different mass ratios. The effects of the heating rate on the reaction rates were determined and key Arrhenius kinetic parameters, i.e., activation energy and pre-exponential factor, were determined. By obtaining thermal loss data from thermogravimetric analysis [14], these kinetic parameters were determined by using pre-determined non-isothermal models such as the Kissinger model, the Flynn-Wall-Ozawa model (FWO), and the improved Kissinger-Akahira-Sunose model (KAS) derived from the generalized Arrhenius model. Results from these methods are compared.

2. Results and Discussion

2.1. Effect of Heating Rate on Conversion and Reaction Rate

Heating rates were varied for date palm biomass at 5 K/min, 10 K/min and 15 K/min to observe their effects on peak temperatures and also to obtain isoconversional data points for kinetic analysis. The mass loss curve for the two biomasses and co-pyrolysis are shown in Figure 1. Each mass loss curve (α) in Figure 1 is the resulting average of triplicates. The mass loss (α) is explained in Section 3.

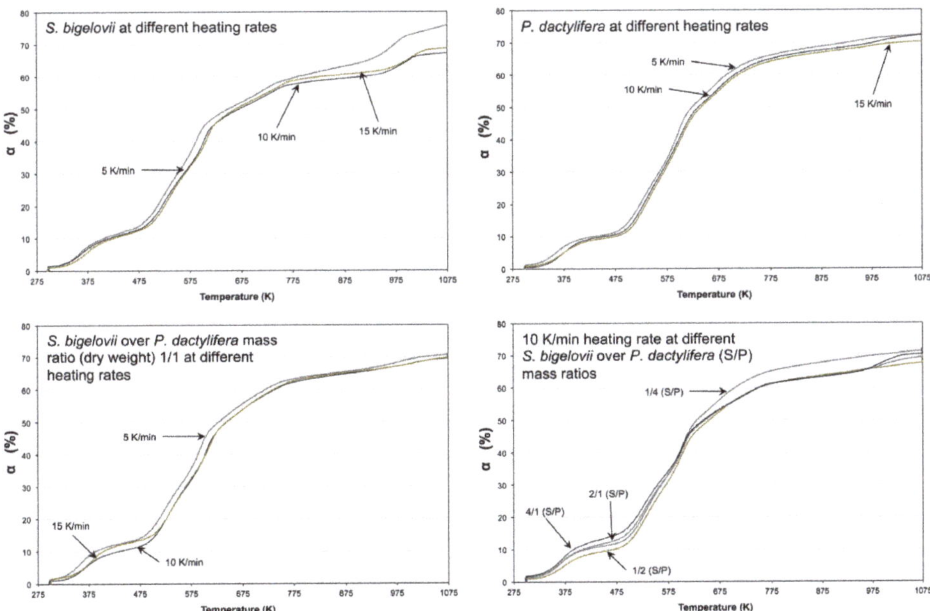

Figure 1. Comparison of the mass loss (α) curves of *S. bigelovii*, *P. dactylifera*, and mixtures of them, at different heating rates (K/min).

The mass loss curves for both biomasses show pyrolysis took place through an identical pathway. Co-pyrolysis biomass followed a pathway between the two pure biomasses as might be predicted; theoretical and kinetic parameters lie between those of the two pure biomasses. The pathway can be categorized into three main phases: Evaporation of water, passive pyrolysis and active pyrolysis which corresponds to the first two phases having characteristic peaks associated with them on the DTG diagram in Figure 2. The evaporation of water occurs around 373 K, corresponding to the first peak on the curve. Active pyrolysis was observed to take place between 473 K and 633 K as seen with the two peaks in this region. Passive pyrolysis began after active pyrolysis and continued till the end of mass loss.

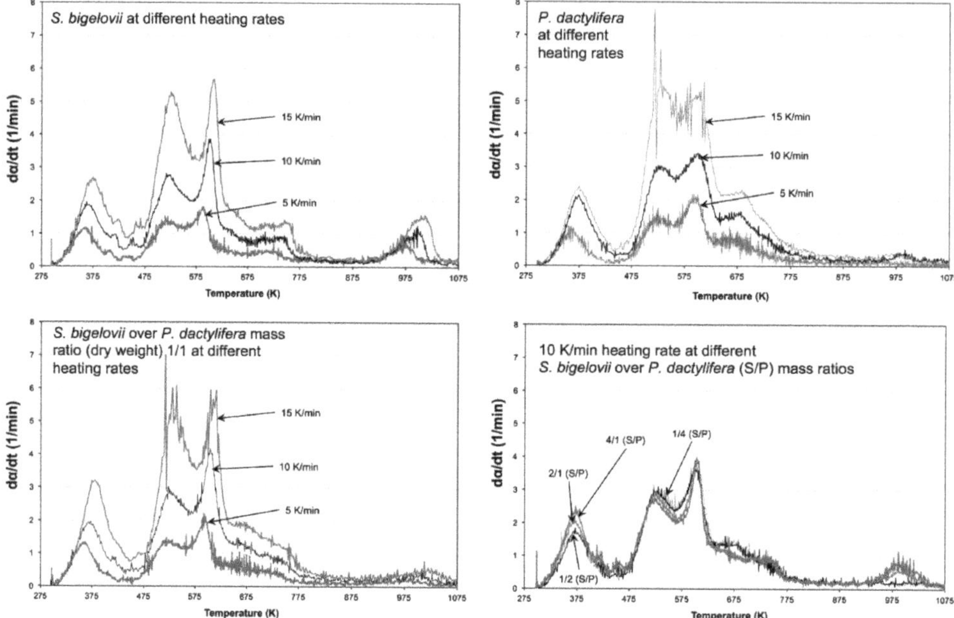

Figure 2. Decomposition rates ($d\alpha/dt$) of *S. bigelovii*, *P. dactylifera*, and mixtures of them, at different heating rates (K/min) showing peak temperatures.

Gasparovic et al. [15] studied the decomposition of hemicellulose, cellulose and lignin and concluded that the decomposition of these three components typically occurred at temperature ranges between 473 K to 653 K, 523 K to 653 K and 453 K up to 1075 K, respectively. This observation reveals that the decomposition of all hemicellulose and cellulose took place during active pyrolysis but decomposition of lignin took place in both active and passive pyrolysis phases.

Increasing heating rate increases the rate of reaction but does not significantly affect the conversion yields at the end of the experiment. Lower heating rates for *S. bigelovii* actually increased the total conversion at the end of the process although the rate of reaction was slower. The reaction rate for all heating rates peaked between 588 K and 602 K during the active pyrolysis phase. These peak temperatures were employed in the kinetic parameter determination using the Kissinger method.

From Figure 2, it is clear that the rate of reaction at all phases of pyrolysis increased with increasing heating rates. However, increased heating rates also led to a more non-uniform reaction decomposition process during the active pyrolysis phase. From the diagram, it can be seen that though the rate of decomposition generally increases with the heating rate and during the active pyrolysis phase, there are no distinct peaks.

2.2. Kinetic Analysis

2.2.1. Kinetic Analysis-*S. bigelovii*

Kinetic parameters were determined for different conversion rates corresponding to different segments of the pyrolysis reaction and the activation energy and pre-exponential factors were determined. The three utilized methods, i.e., Kissinger, FWO, and KAS, are explained in Section 3.

Kissinger Method

The peak temperatures of the DTG curves for *S. bigelovii* and *P. dactylifera* at heating rates of 5 K/min, 10 K/min and 15 K/min were plotted according to the Kissinger model as shown in Figure 3. From the slope of the linear relationship, the calculated activation energy, E (kJ/mol) and the pre-exponential factor, A (1/min), were 147.6 kJ/mol and 3.13×10^9 (1/min) for *S. bigelovii*, respectively, and 164.7 kJ/mol and 9.55×10^{10} (1/min) for *P. dactylifera*, respectively. The peak temperature T_p (K) for *S. bigelovii* at 5, 10 and 15 K/min was 589, 602 and 609 K, respectively. For *P. dactylifera* it was 592, 603 and 612 K.

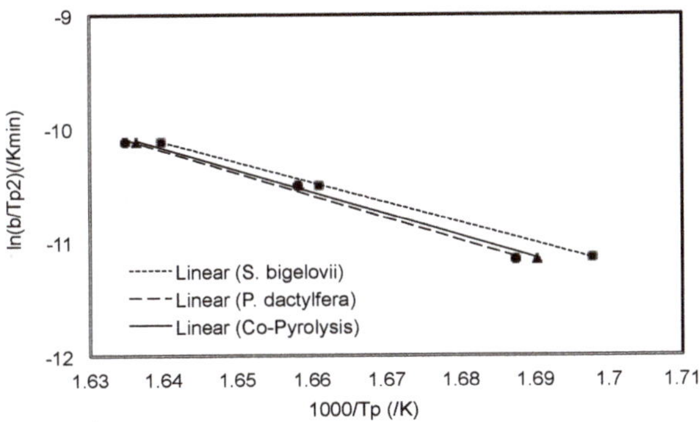

Figure 3. Kissinger curve for *S. bigelovii*, *P. dactylifera* and co-pyrolysis of them at a mass ratio of 1/1, at different isoconversional points.

FWO Method

A plot of $\ln b$ against $\frac{1000}{T_i}$ at the same conversions for heating rates 5 K/min, 10 K/min and 15 K/min was made for varying degrees of conversion of *S. bigelovii* and *P. dactylifera*.

The calculated activation energy, E (kJ/mol) and the pre-exponential factor, A (1/min), were 146.7 kJ/mol and 2.92×10^{20} (1/min) for *S. bigelovii*, respectively, and 204.3 kJ/mol and 1.93×10^{37} (1/min) for *P. dactylifera*, respectively. The FWO plot is shown in Figure 4.

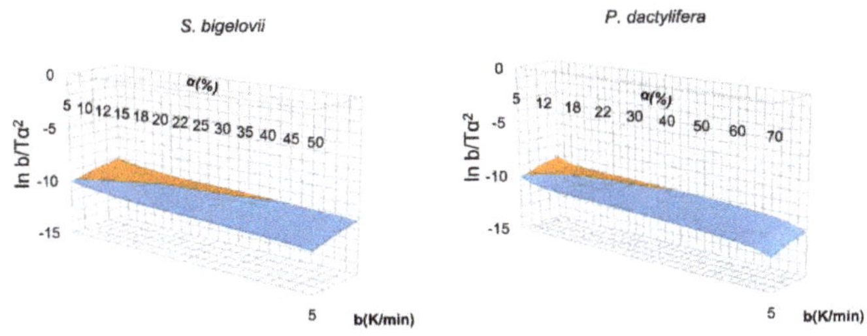

Figure 4. FWO Plot for *S. bigelovii*, *P. dactylifera*.

KAS Method

Figure 5 shows a plot of $\ln \frac{b}{T_\alpha^2}$ against $\frac{1000}{T_\alpha}$ at different heating rates 5 K/min, 10 K/min and 15 K/min at the same conversion for varying conversions from 5% to 50%, during which active pyrolysis takes in *S. bigelovii* and *P. dactylifera*. The calculated activation energy, E (kJ/mol), and the pre-exponential factor, A (1/min), were 147.3 kJ/mol and 3.57×10^{14} (1/min) for *S. bigelovii*, respectively, and 201.4 kJ/mol and 8.54×10^{31} (1/min) for *P. dactylifera*, respectively.

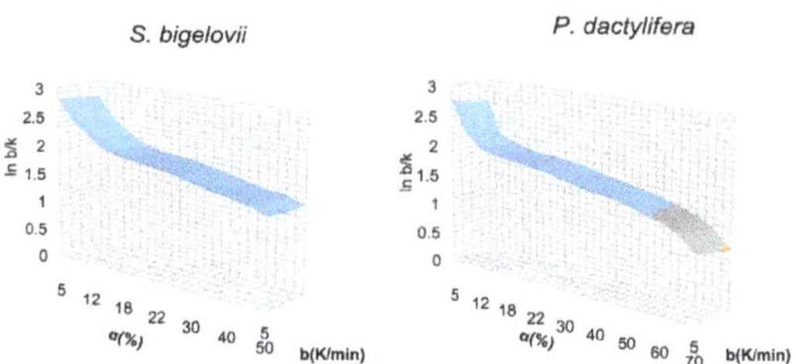

Figure 5. KAS Plot for *S. bigelovii*, *P. dactylifera*.

2.2.2. Summary of Parameters

The kinetic parameters obtained for *S. bigelovii* and *P. dactylifera* leaves from all three methods are summarized in Table 1. The Kissinger method gives one value for the activation energy for the whole process. This value tends to correspond more closely with conversion near peak values at which the Kissinger model was developed. Kinetic parameters for the FWO and KAS method vary with conversion with low values of activation energy at stages prior to active pyrolysis. Activation energy rises during the active phases but tends to reduce at higher conversions during the passive phase for both KAS and FWO methods.

There is very scarce or no literature study on the kinetic parameters of these two biomasses using these methods to compare. Sait et al. [9] have reported activation energies below 44 kJ mol^{-1} for date palm biomass, which significantly deviates from the values reported here, 146–204 kJ mol^{-1}. The activation energy values reported here are in the same order of what has been reported for other non-arid-land lignocellulosic biomass [15–17]. Gasparovic et al. [15] determined the kinetic parameters of wood chip using the generalized isoconversional method and activation energy values of between 131.56 and 215.94 kJ mol^{-1} depending on the conversion. Kongkaew et al. [17] also determined the kinetic parameters for pyrolysis of rice straw using Kissinger, FWO and and KAS methods and obtained 172.62 kJ mol^{-1} for the Kissinger method and activation energy values between 180.54 to 220.27 kJ mol^{-1} and 181.95 to 221.72 kJ mol^{-1} for FWO and KAS, respectively [17]. The same author also obtained 1.46×10^{11} min^{-1} as the pre-exponential factor using the Kissinger method as well. Bartocci et al. [16] have reported activation energies for the three main components of lignocellulosic biomass, 154.1 kJ mol^{-1}, 224.7 kJ mol^{-1}, and 190.5 kJ mol^{-1} for hemicellulose, cellulose, and lignin, respectively.

This shows the activation energy values obtained for both *P. dactylifera* leaves and *S. bigelovii* are within the range of values obtained by other authors using other non-arid-land lignocellulosic biomasses as reactants. Since the Kissinger method adopts the same method to obtain an average value of activation during active pyrolysis, these values are more comparable with values obtained from other authors using different biomasses. FWO and KAS methods, though generally more accurate than the Kissinger method [18], tend to cite averages which also depend on data points. These estimates, however, help to understand the pyrolysis reaction and calculate reaction constants at various stages of the pyrolysis. Also, the fact that the present results are similar to those obtained for non-arid-land lignocellulsic biomass, has interesting implications, e.g., the scale up of the studied arid-land biomass can benefit from the non-arid-land technology.

Table 1. Sumary of kinetic parameters E (kJ/mol) and A (1/min) for *S. bigelovii*, *P. dactylifera* by different methodologies.

Method	*S. bigelovii*		*P. dactylifera*	
	E **(KJ/mol)**	A **(min^{-1})**	E **(KJ/mol)**	A **(min^{-1})**
Kissinger	147.6	3.13×10^9	164.7	9.55×10^{10}
FWO	146.7	2.92×10^{20}	204.3	1.93×10^{37}
KAS	147.3	3.57×10^{14}	201.4	8.54×10^{31}

3. Materials and Methods

Dried *S. bigelovii* (whole plant) were obtained from ISEAS farms in Abu Dhabi. Dried *P. dactylifera* leaves were also obtained from a farm in Abu Dhabi. All feedstock were shredded, milled and filtered through a sieve to obtain particle sizes below 0.5 mm. A sample of *S. bigelovii* and *P. dactylifera* were milled together in the ratio 1:1 for co-pyrolysis. Thermogravimetric analysis was conducted with a Netzch STA 449F3 STA449F3A-0625-M instrument and an aluminum crucible (Al_2O_3) under a Nitrogen atmosphere. The mass loss of biomass (α) was observed over a temperature program from 283 K to 1073 K. For each test, the mass loss (α) was performed in triplicates. The heating rate was kept constant at 10 K/min. Isothermal conditions were created at the start and end of the temperature program to eliminate noise. The heating rate was then varied at 15 K/min and 5 K/min respectively for both biomasses.

3.1. Isoconversional Methods for Kinetic Parameter Estimation-Theory

The pyrolytic reaction of biomass converts biomass to char, oil and gases. The kinetics of the reaction can be described by defining a degree of conversion and using the Arrhenius equation. The isoconversional method is based on the fact that activation energy and pre-exponential are not constant throughout the decomposition but depend on the degree of conversion. Data points at the same conversion are gathered for different heating rates and each isoconversional curve is used to estimate the kinetic parameters at that conversion.

If m_i is the initial mass of sample placed in the crucible, and m_f is the mass of sample left after pyrolysis, a degree of conversion (α) for a given sample mass m at any temperature T during the process can be defined as:

$$\alpha = \frac{m_i - m}{m_i - m_f} \tag{1}$$

From the Arrhenius equation,

$$k(T) = Ae^{\frac{-E}{RT}} \tag{2}$$

where k is the reaction rate constant (varies with temperature), E is the activation energy, A is the pre-exponential factor, R is the gas constant, and T is temperature in Kelvin.

The rate of decomposition can then be defined as a function of conversion and temperature:

$$\frac{d\alpha}{dt} = f(\alpha)k(T)$$

$f(\alpha)$ can be expressed as:

$$f(\alpha) = (1 - \alpha)^n,$$

where n is the reaction order.

Since the temperature program was run with a constant heating rate from 298 K to 1073 K, the temperature at any time t can be written as:

$$T = Ti + bt$$

where T = initial temperature = 298 K and b is the constant heating rate.

The rate of decomposition can be written as:

$$\frac{d\alpha}{dt} = (1 - \alpha)^n A e^{\frac{-E}{RT}} \tag{3}$$

3.1.1. Kissinger Method

For a predetermined reaction order, the rate of decomposition can be plotted at the same conversion for different heating rates. Also the decomposition rate is maximum at peak temperatures (T_p).

$$\frac{d}{dt}\frac{d\alpha}{dt} = \frac{d}{dt}\left((1 - \alpha)^n A e^{\frac{-E}{RT}}\right) = 0 \tag{4}$$

$$\ln \frac{b}{T_p^2} = -\frac{E}{RT_p} + \ln \frac{AR}{E} \tag{5}$$

A plot of $\ln \frac{b}{T_p^2}$ against $\frac{1000}{T_p}$ gives a straight line with slope $-\frac{E}{R}$.

3.1.2. FWO Method

Flynn-Wall-Ozawa developed a method where the activation energy is found by plotting the heating rates against the temperature at which a given conversion is obtained at that heating rate.

$$\ln b = -1.052 \frac{E_\alpha}{RT_\alpha} + \ln \frac{A_\alpha E_\alpha}{R f(\alpha)} - 5.331 \tag{6}$$

A plot of $\ln b$ against $\frac{1000}{T_i}$ gives a straight line with slope $-1.052 \frac{E_\alpha}{R}$. This is used to find E_α.

3.1.3. KAS Method

The Kissinger-Akahira-Sunnose Method

$$\ln \frac{b}{T_\alpha^2} = -\frac{E_\alpha}{RT_\alpha} + \ln \frac{A_\alpha R}{E_\alpha f(\alpha)} \tag{7}$$

A plot of $\ln \frac{b}{T_\alpha^2}$ against $\frac{1000}{T_\alpha}$ at different heating rates at the same conversion gives a straight line with slope $-\frac{E_\alpha}{R}$. The activation energy at that conversion E_α is found from the slope.

4. Conclusions

Activation energy obtained using the Kissinger method for *S. bigelovii* and *P. dactylifera* leaves were 147.6 KJ/mol and 164.7 KJ/mol, respectively, while pre-exponential factors of 3.13×10^9/min and 9.55×10^{10}/min for *S. bigelovii* and *P. dactylifera* leaves were obtained from DTG data. Other values at different stages during the pyrolysis process were also obtained using other isoconversional methods like the FWO, KAS and the generalized isoconversional theories. The mass loss data also revealed the peak temperatures at which reaction is fastest during active pyrolysis. Results of DTG also showed the reaction proceeded in similar phases for the two sampled biomass types, and co-pyrolysis of them at different mass ratios.

Author Contributions: Conceptualization, J.-R.B.-O. and J.E.S.; Methodology, P.D. and J.-R.B.-O.; Formal Analysis, P.D. and J.-R.B.-O.; Investigation, P.D.; Resources, J.-R.B.-O.; Writing-Original Draft Preparation, P.D.; Writing-Review & Editing, J.-R.B.-O. and J.E.S.; Supervision, J.-R.B.-O. and J.E.S.; Project Administration, J.-R.B.-O. and J.E.S.; Funding Acquisition, J.E.S.

Funding: This research was funded by Masdar Institute, Project 2BIONRG (12KAMA4) and BIOREF (13KAMA1), to help fulfil the vision of the late President Sheikh Zayed Bin Sultan Al Nahyan for sustainable development and empowerment of the United Arab Emirates and humankind.

Acknowledgments: Special thanks goes to ISEAS for supplying the biomasses needed for the project.

Conflicts of Interest: The authors declare no conflict of interest.

References

1. Bastidas-Oyanedel, J.R.; Fang, C.; Almardeai, S.; Javid, U.; Yousuf, A.; Schmidt, J.E. Waste biorefinery in arid/semi-arid regions. *Bioresour. Technol.* **2016**, *215*, 21–28. [CrossRef] [PubMed]
2. SBRC Pilot Facility Integrated Seawater Energy and Agriculture System. Available online: https://sbrc.masdar.ac.ae/index.php/projects/seas/item/76-the-seawater-energy-and-agriculture-system (accessed on 20 June 2011).
3. Isahak, W.N.R.W.; Hisham, M.W.M.; Yarmo, M.A.; Hin, T.-Y.Y. A review on bio-oil production from biomass by using pyrolysis method. *Renew. Sustain. Energy Rev.* **2012**, *16*, 5910. [CrossRef]
4. Czernik, S.; Scahill, J.; Diebold, J. The Production of Liquid Fuel by Fast Pyrolysis of Biomass. *J. Sol. Energy Eng.* **1995**, *117*, 2. [CrossRef]
5. Bridgwater, A.V. Review of fast pyrolysis of biomass and product upgrading. *Biomass Bioenergy* **2012**, *38*, 68–94. [CrossRef]
6. Demirbas, A. Determination of calorific values of bio-chars and pyro-oils from pyrolysis of beech trunkbarks. *J. Anal. Appl. Pyrolysis* **2004**, *72*, 215–219. [CrossRef]
7. Uçar, S.; Karagöz, S. The slow pyrolysis of pomegranate seeds: The effect of temperature on the product yields and bio-oil properties. *J. Anal. Appl. Pyrolysis* **2009**, *84*, 151–156. [CrossRef]
8. Solar, J.; de Marco, I.; Caballero, B.M.; Lopez-Urionabarrenechea, A.; Rodriguez, N.; Agirre, I.; Adrados, A. Influence of temperature and residence time in the pyrolysis of woody biomass waste in a continuous screw reactor. *Biomass Bioenergy* **2016**, *95*, 416–423. [CrossRef]
9. Sait, H.H.; Hussain, A.; Salema, A.A.; Ani, F.N. Pyrolysis and combustion kinetics of date palm biomass using thermogravimetric analysis. *Bioresour. Technol.* **2012**, *118*, 382–389. [CrossRef] [PubMed]
10. Conti, L.; Scano, G.; Boufala, J. Bio-oils from arid land plants: Flash pyrolysis of Euphorbia characias bagasse. *Biomass Bioenergy* **1994**, *7*, 291–296. [CrossRef]
11. Putun, A.; Gercel, H.; Kockar, O.; Ege, O.; Snape, C.; Putun, E. Oil production from an arid-land plant fixed bed pyrolysis and hydropyrolysis. *Fuel* **1996**, *75*, 1307–1312. [CrossRef]
12. Warshay, B.; Brown, J.J.; Sgouridis, S. Life cycle assessment of integrated seawater agriculture in the Arabian (Persian) Gulf as a potential food and aviation biofuel resource. *Int. J. Life Cycle Assess.* **2017**, *22*, 1033. [CrossRef]
13. Abideen, Z.; Ansari, R.; Gul, B.; Khan, M.A. The place of halophytes in Pakistan's biofuel industry. *Biofuels* **2012**, *3*, 211–220. [CrossRef]
14. Buratti, C.; Barbanera, M.; Bartocci, P.; Fantozzi, F. Thermogravimetric analysis of the behavior of sub-bituminous coal and cellulosic ethanol residue during co-combustion. *Bioresour. Technol.* **2015**, *186*, 154–162. [CrossRef] [PubMed]
15. Gasparovic, L.; Korevnova, Z.; Jelemensky, L. Kinetic study of wood chips decomposition by TGA. *Chem. Pap.* **2010**, *64*, 174–181. [CrossRef]
16. Bartocci, P.; Anca-Couce, A.; Slopiecka, K.; Nefkens, S.; Evic, N.; Retschitzegger, S.; Barbanera, M.; Buratti, C.; Cotana, F.; Bidini, G.; et al. Pyrolysis of pellets made with biomass and glycerol: Kinetic analysis and evolved gas analysis. *Biomass Bioenergy* **2017**, *97*, 11–19. [CrossRef]
17. Kongkaew, N.; Pruksakit, W.; Patumsawad, S. Thermogravimetric Kinetic Analysis of the Pyrolysis of Rice Straw. *Energy Procedia* **2015**, *79*, 663–670. [CrossRef]
18. Starink, M.J. The determination of activation energy from linear heating rate experiments: A comparison of the accuracy of isoconversion methods. *Thermochim. Acta* **2003**, *404*, 163–176. [CrossRef]

MDPI
St. Alban-Anlage 66
4052 Basel
Switzerland
Tel. +41 61 683 77 34
Fax +41 61 302 89 18
www.mdpi.com

Energies Editorial Office
E-mail: energies@mdpi.com
www.mdpi.com/journal/energies